T0332942

# Artificial Intelligence in Mechanical and Industrial Engineering

**Artificial Intelligence (AI) in Engineering**
Series Editors: Kaushik Kumar and J. Paulo Davim

This new series will target and provide a collection of textbooks and research books on Artificial Intelligence (AI) applied in engineering disciplines. The series of books will provide an understanding of AI using common language and incorporate tools and applications to assistant in the learning process. They will focus on areas such as Artificial Intelligence and Philosophy, Applications of AI in Mechatronics, AI in Automation, AI in Manufacturing, AI and Industry 4.0, Cognitive Aspects of AI, Intelligent Robotics, Smart Robots, COBOTS, Machine Learning, Conscious Computers, Intelligent Machines, to name just a few areas that the series will have books in.

**Nature-Inspired Optimization in Advanced Manufacturing Processes and Systems**
*Edited by Ganesh M. Kakandikar and Dinesh G. Thakur*

**Watershed Management and Applications of AI**
*Sandeep Samantaray, Abinash Sahoo, and Dillip K. Ghose*

**Artificial Intelligence in Mechanical and Industrial Engineering**
*Edited by Kaushik Kumar, Divya Zindani, and J. Paulo Davim*

For more information on this series, please visit: www.routledge.com/Artificial-Intelligence-AI-in-Engineering/book-series/CRCAIIE

# Artificial Intelligence in Mechanical and Industrial Engineering

**Edited by**

*Kaushik Kumar, Divya Zindani and J. Paulo Davim*

CRC Press
Taylor & Francis Group
Boca Raton London New York

CRC Press is an imprint of the
Taylor & Francis Group, an **informa** business

First edition published 2021
by CRC Press
6000 Broken Sound Parkway NW, Suite 300, Boca Raton, FL 33487-2742
and by CRC Press
2 Park Square, Milton Park, Abingdon, Oxon, OX14 4RN



**Library of Congress Cataloging-in-Publication Data**

Names: Kumar, K. (Kaushik), 1968- editor. | Zindani, Divya, 1989- editor. | Davim, J. Paulo, editor.
Title: Artificial intelligence in mechanical and industrial engineering / edited by Kaushik Kumar, Divya Zindani and J. Paulo Davim.
Description: First edition. | Boca Raton : CRC Press, 2021. | Series: Artificial intelligence (AI) in engineering | Includes bibliographical references and index.
Identifiers: LCCN 2020057786 (print) | LCCN 2020057787 (ebook) | ISBN 9780367441760 (hbk) | ISBN 9781003011248 (ebk)
Subjects: LCSH: Artificial intelligence—Industrial applications. | Mechanical engineering—Data processing. | Industrial engineering—Data processing.
Classification: LCC TA347.A78 .A84 2021 (print) | LCC TA347.A78 (ebook) | DDC 620.00285/63—dc23
LC record available at https://lccn.loc.gov/2020057786
LC ebook record available at https://lccn.loc.gov/2020057787

ISBN: 978-0-367-44176-0 (hbk)
ISBN: 978-1-032-01296-4 (pbk)
ISBN: 978-1-003-01124-8 (ebk)

Typeset in Times LT Std
by KnowledgeWorks Global Ltd.

# Contents

## *SECTION III: Application towards Industrial Engineering*

# Preface

The editors are pleased to present the book *Artificial Intelligence in Mechanical and Industrial Engineering* as a part of the *Artificial Intelligence (AI) in Engineering* series. The book title was chosen looking at the current and growing importance of *artificial intelligence* as well as for familiarization with application of the same to the most important domain, *mechanical and industrial engineering*, for industrial and manufacturing world.

Computer science defines artificial intelligence (AI) research as the study of *intelligent agents*: any device that perceives its environment and takes actions that maximize its chance of successfully achieving its goals. A more elaborate definition characterizes AI as *a system's ability to correctly interpret external data, to learn from such data, and to use those learnings to achieve specific goals and tasks through flexible adaptation*. Hence, AI is the simulation of the process of data interaction of human thinking, hoping to understand the essence of human intelligence and then produce a smart machine. This intelligent machine can be the same as human thinking to respond and deal with the problem. Although thought-capable artificial beings appeared as storytelling devices in antiquity, and have been common in fiction, as in Mary Shelley's *Frankenstein* or Karel Čapek's *R.U.R*, the current field of AI research was born at a workshop at Dartmouth College in 1956, where the term 'artificial intelligence' was coined by John McCarthy to distinguish the field from cybernetics and escape the influence of the cyberneticist Norbert Wiener. Attendees Allen Newell and Herbert Simon (Carnegie Mellon University), John McCarthy and Marvin Minsky (Massachusetts Institute of Technology) and Arthur Samuel (IBM) became the founders and leaders of AI research.

With the continuous progress of science and technology, mechanical engineering is also constantly evolving and changing, from the traditional mechanical engineering to the electronic mechanical engineering. With the continuous progress of science and technology, mechanical engineering is also constantly evolving and changing, from the traditional mechanical engineering to the electronic mechanical engineering. It went into a new stage of development and its level of automation and intellectualization has continuously improved. Thus, the combination of artificial intelligence technology and mechanical and industrial engineering has become a hotspot. Some of the trusted domains in the field of mechanical and industrial engineering which are directly benefitting the society and common man are:

*Quality control*: A common use of AI in manufacturing is for machines to visually inspect items on a production line. Using AI allows quality control to be automated and ensures that all final products are inspected, allowing fewer defects to reach customers compared to traditional statistical sampling methods. In addition to ensuring that products are free of imperfections, AI-based visual inspection systems can validate many product attributes including geometry and tolerances, surface finish, product classification, packaging, colour and texture.

*Fault detection and isolation*: In regulated manufacturing environments, ensuring process compliance can be expensive and time-consuming. In many such scenarios, lives are at stake – as can be the case in food, chemical and energy industries. By monitoring a variety of system operational factors, AI can be used in the detection, prediction and diagnosis of undesirable operating conditions in industrial systems. By accelerating or replacing unreliable and time-consuming human analysis, automated process surveillance helps prevent or minimize system downtime and the persistence of hazardous conditions.

*Anticipatory logistics and supply chain management*: Supply chain management is traditionally a two-step process. First, statistical tools are used to produce a demand forecast. The forecast is then used as input to an optimization process that evaluates the cost of stockouts against the delivery times, holding costs and other factors associated with the supply chain. Supply chain managers can then use tools to produce a plan for what to order and when. Using machine learning, it is now possible to implement a single-step process that learns the relationship between all available input data, including traditional supply chain data such as inventory levels, product orders and competitive data, as well as external data such as weather, social media signals and more, to produce better operational performance.

The objective of this edited book is to provide a unified platform for dissemination of basic and applied knowledge about integration of artificial intelligence within the realm of mechanical and industrial engineering. The specific mission of the book is to give scientists, researchers, teachers, students and practitioners the tools and information they need to build successful careers and also to be an impetus to advancements in next-generation mechanical and industrial domain. The general objective of the book is to become a source of basic and applied knowledge in the field of mechanical and industrial engineering by providing high-quality scientific, engineering and technological information.

The entire book is divided into three sections containing eight chapters. The three sections are Section I: State of Art, Section II: Application towards Mechanical Engineering and Section III: Application towards Industrial Engineering. Section I contains Chapter 1, Section II comprises Chapters 2 to 6 and Section III is composed of Chapters 7 and 8.

Chapter 1 provides the readers with an overview of machine learning implementation in various industrial scenarios, e.g., agriculture, healthcare and enterprise. Primarily with the introduction of Industry 4.0, advancements in IT tools such as Internet of things (IOT), cloud computing, cyber-physical systems, artificial intelligence and virtual reality have popped up a new dimension in industries that incorporates intelligent human-to-machine and machine-to-machine systems. Machine learning (ML) proved to be an effective means to mimic various complex situations and processes involved in product manufacturing, manufacturer–supplier relation and manufacturer–consumer relations and to incorporate them into complicated decision-making algorithms. These incorporations, in turn, help industries in building energy- and time-efficient systems for a better utilization of resources, less involvement of humans and higher revenue creation. In machine learning, the machines are made to self-learn through experiences without being programmed explicitly. In this chapter, a detailed review has been presented for the implementation of ML in various industrial scenarios such as in healthcare industries, agricultural supply chain systems and enterprises.

In Chapter 2, the application of artificial intelligence to predict the process parameters for optimizing wear behaviour of sintered titanium grade 5 (Ti-Al-V) reinforced with nano $B_4C$ particles is discussed. In this work, full factorial design was utilized for the same. The composite samples were prepared using powder metallurgy technique, and using pin-on-disc machine the elevated temperature wear behaviour was analyzed. It was observed that inclusion of nano $B_4C$ drastically reduced the wear irrespective of temperature and sliding distances.

In Section II, Chapter 3, similar to the previous chapter, provides the reader with an exclusive study of tribological behaviour of AL7068-alumina-$B_4C$ hybrid composites and optimization with DEMATEL technique. In this chapter, boron carbide ($B_4C$) and aluminium oxide ($Al_2O_3$) are used as reinforcement. Boron carbide possesses high hardness value, good chemical stability, low specific gravity and high elastic modulus, making it a better reinforcing agent in a matrix material. The influence of boron carbide, as well as aluminium reinforcement, on aluminium metal matrix composite is analyzed in this chapter.

Chapter 4 aims to compare the ant colony optimization (ACO) algorithm and genetic algorithm (GA) for routing automated guided vehicles (AGVs). AGVs are very crucial for transporting materials and products in manufacturing systems, and its flexibility and cost-effectiveness are the salient required features. This chapter provides comparison of effectiveness between ACO and GA for routing AGVs by applying algorithm developed using C# programming language. The AGVs' effectiveness got increased and it was observed that ant colony algorithm was more successful for solving the travelling salesman problems (TSP).

Chapter 5 discusses the work done in designing and controlling of a two-link robotic manipulator along with model-based controller design using conventional PID-type control and computed torque control. These conventional control methods are associated with many flaws such as variational parameters, variational payloads, backlashes and uncertainties, but still these are much widely used in industrial applications due to their ease of control and design and lower costs. Hence, for better performance and desired results, these methods cannot be relied upon and the time and advancements in technologies have forced to adopt the alternatives for new kinds of applications along with their associated complications. The chapter introduces mainly two types of intelligent controlling approach, fuzzy logic control (FLC) and artificial neural network (ANN) control, which are model-free approach and much popular and have easy adaptive properties with known complexities. Subsequently, the advantages and disadvantages for each controller have been discussed.

In Chapter 6, the last chapter of Section II, deep learning was utilized to detect and classify maize leaf disease. Deep learning models are extensively used in precision farming and smart agriculture. In this chapter, different maize leaf diseases, such as grey leaf spot, common rust and northern leaf blight, were analyzed using deep learning techniques. Moreover, two convolutional neural network (CNN) models, namely, custom-designed CNN and a pre-trained neural network called AlexNet with transfer learning, were used for classification of maize leaf diseases. Classification results were critically evaluated using different multiclass classification metrics. The results are competent with the existing methods in this domain.

The proposed material and models were able to give promising classification with high-accuracy results.

Chapter 7, the first chapter of the last section, i.e., Section III, focuses on the development of new model based on the natural constraints, which is inspired by the human cognition process that could reason the differences in patterns and recognize them to solve problems. Partial usage of network topology has been applied to identify cells and their index, with over five differential sub-methods devised to work adjacent to the main module, which interlinked with each other to find the patterns. By using the IBC algorithm, it has been observed that when this model is applied to large-sized machine component matrix, the time and solution search have lowered. On comparing with the benchmark problem sets, the solution time as the search traversal has greatly decreased, retaining the same optimal solution.

Chapter 8, the last chapter of Section III and of the book, presents optimization of operating parameters of a non-traditional machining process, namely, wire EDM. The requirement of alloy materials with great hardness, strength and impact tolerance has been raised by the growth of the automotive sector to satisfy diverse needs. Therefore, non-traditional machining techniques, such as electrochemical, ultrasonic and electrical discharge machining (EDM), are employed to machine materials that are too difficult to machine. Wire electric discharge machine (WEDM), which utilizes a wire electrode to initiate the sparking mechanism, is a special form of the traditional EDM process. In this chapter, cryogenic-treated D2 steel and cryogenic-treated Bronco cut wire have been used, which is absolutely new. Hence, a comparative study of optimum parameters for cryogenic-treated and normal wire and workpiece was essential for validating and comparing the results and scope of cryogenic treatment in WEDM. Taguchi-based optimization was used in this work for finding optimum process parameters for machining of D2 steel for lower surface roughness (SR). The surface roughness got substantial reduction, validating the usage.

First and foremost, we would like to thank God. It was your blessing that provided us the strength to believe in passion and hard work and to pursue our dreams. We also thank our families for having the patience with us for taking yet another challenge which decreases the amount of time we could spend with them. They were our inspiration and motivation. We would also like to thank our parents and grandparents for allowing us to follow our ambitions. We would like to express our gratitude to all the contributing authors, as they are the pillars of this structure. We thank them for believing in us. We are grateful to all of our colleagues and friends in different part of the world for sharing their ideas in shaping our thoughts. Our efforts will come to a level of satisfaction if the students, researchers and professionals concerned with all the fields related to nanomaterials and nanocomposites, in particular, and material science and product development, in general, benefit from our work.

We owe a huge thanks to each and every contributing authors, reviewers, editorial advisory board members, development editor and the team of CRC Press for their availability for work on this huge project. All of their efforts were instrumental in compiling this book, and without their constant and consistent guidance, support and cooperation, we could not have reached this milestone, especially in this pandemic situation.

Last, but definitely not least, we would like to thank all the individuals who had taken time out and help us during the process of writing this book. Without their support and encouragement, we would have probably given up the project.

**Kaushik Kumar**
*Birla Institute of Technology, Mesra*

**Divya Zindani**
*Sri Sivasubramaniya Nadar (SSN) College of Engineering, Kalavakkam*

**J. Paulo Davim**
*University of Aveiro*

# Editors

**Kaushik Kumar,** BTech (Mechanical Engineering, REC [now NIT], Warangal), MBA (Marketing, IGNOU) and PhD (Engineering, Jadavpur University), is presently associate professor, Department of Mechanical Engineering, Birla Institute of Technology, Mesra, Ranchi, India. He has 19 years of experience in teaching and research and over 11 years of industrial experience in a manufacturing unit of global repute. His areas of teaching and research interest are composites, optimization, non-conventional machining, CAD/CAM, rapid prototyping and quality management systems. He has 9 patents, 35+ books, 30+ edited books, 55+ book chapters, 150+ international journal publications and 22 international and 1 national conference publications to his credit. He is on the editorial board and review panel of seven international and one national journals of repute. He has been felicitated with many awards and honours (Web of Science core collection [104 publications/h-index 10+, SCOPUS/h-index 10+, Google Scholar/h-index 23+]).

**Divya Zindani,** BE (Mechanical Engineering, Rajasthan Technical University, Kota) and ME (Design of Mechanical Equipment, BIT Mesra), Ph.D. (National Institute of Technology Silchar, thesis submitted). He is currently an assistant professor, Dept. of Mechanical Engineering, Sri Sivasubramaniya Nadar (SSN) College of Engineering. He has over 2 years of industrial experience. His areas of interest are Decision support systems based on AI/ML, Green materials, rapid prototyping and Supply chain. He has 1 patent, 8 authored books, 8 edited books, 23 book chapters, 11 SCI journals, 7 Scopus indexed international journals and 7 international conference publications to his credit.

**J. Paulo Davim** is a full professor at the University of Aveiro, Portugal. He is also a distinguished honorary professor in several universities/colleges in China, India and Spain. He received his PhD degree in Mechanical Engineering in 1997, MSc degree in Mechanical Engineering (Materials and Manufacturing Processes) in 1991, Mechanical Engineering degree (5 years) in 1986 from the University of Porto (FEUP), the Aggregate title (Full Habilitation) from the University of Coimbra in 2005 and the DSc (Higher Doctorate) from London Metropolitan University in 2013. He is senior chartered engineer by the Portuguese Institution of Engineers with an MBA and Specialist titles in Engineering and Industrial Management as well as in Metrology. He is also EUR ING by FEANI-Brussels and fellow (FIET) of IET London. He has more than 30 years' experience in teaching and research in manufacturing, materials, mechanical and industrial engineering, with special emphasis in machining and tribology. He has also interest in management, engineering education and higher education for sustainability. He has guided large numbers of postdoc, PhD and master's students as well as has coordinated and participated in several financed research projects. He has received several scientific awards and honours. He has worked as evaluator of projects for ERC-European Research Council and other international research agencies as well as examiner of PhD thesis for many universities in different countries. He is the editor-in-chief of several international journals, guest editor of journals, books editor, book series editor and scientific advisory for many international journals and conferences. Presently, he is an editorial board member of 30 international journals and acts as reviewer for more than 100 prestigious Web of Science journals. In addition, he has also published as editor (and co-editor) more than 150 books and as author (and co-author) more than 15 books, 100 book chapters and 500 articles in journals and conferences (more than 280 articles in journals indexed in Web of Science core collection/h-index 57+/10500+ citations, SCOPUS/h-index 62+/13000+ citations, Google Scholar/h-index 80+/21500+ citations).

# Section I

## State of Art

# 1 An Overview of Machine Learning Implementation in Various Industrial Scenarios

## *Agriculture, Healthcare and Enterprise*

*Hridayjit Kalita*
Birla Institute of Technology, Mesra
Ranchi, India

*Kaushik Kumar*
Birla Institute of Technology, Mesra
Ranchi, India

*J. Paulo Davim*
University of Aveiro
Aveiro, Portugal

**CONTENTS**

## 1.1 INTRODUCTION

Machine learning (ML) was first introduced about 60 years ago. However, it was not until recently that the system has been able to grab the attention of the industries and businesses in both the economic and social fronts. With the latest innovations and improvements, it has become a key enabler of competitiveness and sustainability and finds its use in healthcare systems, agricultural supply chain systems, enterprises, banking and transportation. Application of ML in these sectors facilitates development of automation and intelligent systems that can solve complex tasks, reduce human efforts, enhance resource utilization, enhance customer experience and encourage customer involvement.

The global extensive use of ML has been triggered mainly by three technological trends: the big data, research and development of ML technologies, and enhanced processing and data storing capabilities of computers. In agricultural sector, the importance of ML algorithms has been stressed in maintaining a sustainable agricultural supply chain (SASC) system by automated monitoring of crops, soil and water with the help of emerging technologies such as Internet of things (IOT), mobile technology, digital delivery of services, data analytics and RFID product traceability [OECD, 2019]. In healthcare systems, the amount of digital data available for the analysis of health conditions of patients is huge, and it has practically become impossible for practitioners to make decision based on reviews of about 80+ GB worth of data from an individual [Brown, 2015] in a short duration. Also, since the healthcare practitioners generally lack ML skills but are excellent in their domain of knowledge, it becomes hard for them to interpret the pattern generated by an ML algorithm, and thus they have to rely on data scientists [Beam and Kohane, 2018;Rajkomar et al., 2019]. One way to avoid such situations is by using AutoML [Quanming et al., 2018], which is an excellent alternative that employs complex algorithms in the form of a black box for a better user-friendly experience, thus enabling healthcare personnel to carry out their models without the need of any ML experts. In enterprises, ML applications have reached its inflexion point, which made it very essential for managers at all levels to adapt to changing technologies and familiarize with ML techniques for better economic, social and environmental performance. In this chapter, Sections 1.2–1.4 give a detailed description of the application of ML algorithms in different operations, their frameworks and types considering agricultural, healthcare and enterprise sectors, respectively.

## 1.2 MACHINE LEARNING IN AGRICULTURE

As the global population rises and is expected to reach 9–10 billion in 2050, shortage of food is inevitable with the current condition of the agricultural supply chain, which must be handled sustainably. ML finds its application in maintaining SASCs by providing information to the farmers on various relating factors and encouraging them to make better decisions and adopt sustainable measures [Adnan et al., 2018]. The fluctuation in the market prices of the agricultural products and the demand-supply gap [Patidar et al., 2018; Sharma et al., 2018] contribute to the main objective of the ML application in agricultural research, along with few unavoidable complications in transportation, weather patterns and inventory issues [Hao et al., 2018; Kazemi et al., 2018; Sarkar and Giri, 2018; Sayyadi and Awasthi, 2018a; Shah et al.,2018; Gharaei et al., 2019b; Gharaei et al., 2019; Hoseini Shekarabi et al., 2019], which must be addressed in the ML framework towards achieving an SASC system. Intelligent systems such as AI or cognitive-based technologies can implement the complex decision-making processes in the SASC by analysing the big data available on the Internet or cloud and manipulating (moving) positions of different objects accordingly. ASC generally involves a number of operations (pre-production, production, distribution, inventory storage) handled by different stakeholders (farmers, processors, final consumers, traders, distributers, retailers, certification agencies, etc.) that requires strategic and tactical management with rapid real-time decision-making capabilities. These complex decision-making problems can be solved with the modern computer and information technologies by exploiting the innumerous data obtained from the sensors and other devices. This adaption can also elevate the understanding of the complex nature of the agricultural ecosystem [Kamilaris et al., 2017] by constantly monitoring the environment around. Monitoring generates a lot of data (both labelled and unlabelled), which are fed to specific ML algorithm as training input data utilizing both previous and current data for improving the prediction or decision-making performance [Du and Sun, 2006]. Different ML algorithms are employed for supervised learning employing labelled data, unsupervised learning employing unlabelled data and reinforced learning employing all data, in addition to the data obtained by interaction with the environment. In ASC, ML algorithms are employed considering the nature of different operations [Sharma et al., 2020] mentioned previously for an efficient ASC system. Each operation in SASC system can be further subdivided, which will be discussed in detail below along with the ML framework.

### 1.2.1 Operational Phases Involved in Agricultural Supply Chain (ASC)

**Pre-production** is the first step towards obtaining a sustainable agricultural practice, which involves factors such as crop yield prediction, predicting the quality of the soil and irrigation requirement of the crops. The crop yield and quality generally depend on the property of the soil, right amount of light and amount of irrigated water. ML algorithm helps in adopting a precision agricultural technique by solving the supply and soil management system including the marketing strategies, crop yield estimation and planning for the requirement of nutrients, fertilizers, soil quality, irrigation and equipment. Soil quality is generally detected by measuring the moisture content [Im et al., 2016; Prasad et al., 2018], nutrient content [Morellos et al., 2016] and soil temperature. Irrigation can be an intelligent system involving the conditions of when, where and how much to irrigate considering the data of precipitation, evaporation and soil moisture content.

**Production** in a SASC system involves multiple factors (such as weather forecasting, nutrient management, disease detection, weed detection, harvesting and crop quality detection), the output condition patterns of which must be analyzed and identified using ML algorithms. Weather forecasting determines the amount of sunlight, rainfall and humidity for planning and scheduling of the irrigation schemes [McNider et al., 2014; Traore et al., 2016]. Maintaining an adequate nutrient content of soil [Sirsat et al., 2018] is vital for a sustainable soil management system, which can be determined by measuring the crop quality parameters, thereby enhancing productivity and minimizing losses [Maione et al., 2016]. Diseases such as pathogen attacks and infestations on crops can be detected even before the actual outbreaks [Lee et al., 2010] by precision agriculture technologies such as site-specific management. Weed detection in crop fields are generally carried out by employing ML combined with machine vision to observe minute differences in their textures, colour and shape from the actual crops [Zhang et al., 2014; Tang et al., 2017]. In deciding allocation of resources for harvesting and post-harvesting strategies [Ahn et al., 2018], crop yield and production estimates are the key variables that are solved using ML algorithms and remote sensing data [Haghverdi et al., 2018].

**Distribution** generally connects the production operation to the final usage by consumers [Manzini et al., 2019] and involves uncertainty factors in the domain of transportation, consumer analytics and inventory management. Transportation factors can be in the form of minimizing the product damage during travel, routine check of the vehicles, determining travel distance, integrating production and distribution scheduling, and preserving the food quality in the process [Qiang and Jiuping, 2008; Rabbani et al., 2016; Buelvas Padilla et al., 2018; Wang et al., 2018]. Consumer analytics involves prediction of consumer demands, their buy behaviour and perception [Ribeiro et al., 2018], with the objective of capturing the attention of the consumers and building a customer-centric atmosphere by getting feedback from various social media platforms.

**Inventory storage** must be managed based on the customer demands to avoid product shortage or out-of-stock conditions, thereby fulfilling customer satisfaction and maintain profitability in the supply chain system.

### 1.2.2 Framework of ML-ASC

The integration of ML algorithms with ASC systems enhances the overall profit, efficiency and environment quality by reducing wastes and commodity losses. The data is generated from various sources (sensors and drones) in different operational phases, as described above, of the ASC system, which pass through different ML algorithms for obtaining context-specific and control capability decisions. The framework can be adopted in three different components, namely ASC phase, ML algorithm and ASC performance.

ASC phase: The data obtained in a single operational phase (e.g. pre-production) corresponding to various factors, such as climate, irrigation, crop yield, and soil, can also be employed for the decision-making in other operational phases (e.g. production), thereby finding and analysing new patterns in the whole supply chain. The focus of the ASC phase is to adopt appropriate technologies for different complex situations and implement them accordingly. IOT platform has the capability to integrate with the drone and sensor technologies for accomplishing sustainable agricultural practices and controlling operational aspects of farms [Porter and Heppelmann, 2015] by forming a smart web of devices connected to the Internet or cloud. Drones have the potential of monitoring crops and managing livestock, while sensors and actuators can help in weed detection, disease detection, pest managements and other environmental and climate conditions.

ML algorithms: ML algorithms can be employed in different operational phases of ASC, depending on the nature and objective of the problem, by incorporating feedback loops between the detectors in the ASC phases and the ML algorithms. ML techniques employed generally come under supervised learning, unsupervised learning and reinforced learning, which gather maximum information from the sources and analyze it for enhancing ASC performance.

SASC performance: A sustainable ASC is the major purpose of integrating ML algorithms with the ASC operational phases. It analyses the agricultural data and look into all economic, social and environmental performances. In the pre-production phase, forecasting rainfall and predicting weather patterns can be employed for effective water and resource planning, harvest planning, crop sowing pattern predicting and avoiding events such as flood and drought. In the production phase, optimized nutrient management and site-specific management are some of the features for enhancing sustainability, which increases crop yield and productivity [Chlingaryan et al., 2018] and reduces food losses, environmental impact and operating costs. In the distribution phase, sustainability in social, environmental and economic aspects [Elkington, 1998] can be achieved by forecasting demand, enhancing safety and quality of food, maintaining economic savings and managing fleet and vehicle routing for fuel savings.

## 1.3 MACHINE LEARNING IN HEALTHCARE

In healthcare systems, the use of digital health devices and wearable electronic health records (EHRs) and genome sequencing have accelerated the rate of obtaining medical data, which can be termed as biomedical 'big data' [Murdoch and Detsky, 2013; Toga et al., 2015; Luo et al., 2016]. In order to make use of such volume of data, ML algorithms and the more recent 'deep learning' can be effectively employed to give a sense of the various patient health patterns to the practitioners and to bring out logical and actionable results. ML algorithms are generally employed in healthcare systems to improve the quality of care given to the patients [Rumsfeld et al., 2016; Kuo et al., 2019; Liang et al., 2019], their safety [Saria et al., 2010; Marella et al., 2017; Lundberg et al., 2018], to reduce healthcare cost [Bates et al., 2014; Özdemir and Barshan, 2014; Lo-Ciganic et al., 2015] and to find its applications in circumstances such as recognizing and transferring the high-risk patients to the ICU [Escobar et al., 2016], early detection of lung cancer [Ardila et al., 2019] and identifying respiratory diseases from chest X-rays [Rajpurkar et al., 2017].

ML has many demonstrated applications in healthcare systems. However, healthcare researchers suffer from the inability to implement ML to their plans, while they are well accustomed to identifying various medical conditions from the available clinical data. In such situations, healthcare researchers generally collaborate and interact with expert data scientists but even then the process demands time and effort from both sides. Such difficulties in deploying ML solutions can be avoided by employing AutoML, which is an excellent alternative to requiring human expertise in computational works, and is able to automate components that optimize data for rapid building, validating and deployment of ML solutions. A black box can be constructed that does not require user expertise and can be applied to problems, thereby empowering the domain of knowledge of the practitioners and coping up with the shortage of data scientists. The AutoML challenge competitions, which were held between 2015 and 2018, focused on solving supervised ML problems with ranges of computational constraints varying in time and memory usage limits across the challenge. The AutoML algorithms in healthcare services can be addressed based on their strategies to automate problems in a healthcare setting. These strategies are automated feature engineering, optimization of hyper-parameters, pipeline optimization and neural architecture [Waring et al., 2020].

### 1.3.1 Automated Feature Engineering

A data scientist is generally expected to create some detailed explanatory variables termed as 'features', using his/her expertise and domain of knowledge, that give an insight into the clinical data. These features are employed for solving various supervised ML problems while identifying ML algorithm limitation, and their qualities are critical to the performance and outcome result of the ML algorithms for any decision makers. The features are determined by trial-and-error method, which is a drawback of feature engineering in terms of consumed time and effort of the decision makers.

Frameworks for automated feature engineering have been developed to tackle the drawbacks of traditional feature engineering, where novel feature sets are decided for enhancing performance of ML algorithms. Representative learning, which employs deep learning to find appropriate feature spaces in the unstructured data type, can be employed in ML pipeline. This cannot be considered as an AutoML technique but is useful in healthcare setups in finding useful representations from EHR data and predict health conditions [Miotto et al., 2016; Rajkomar et al., 2018]. The automated feature engineering tasks generally involve a set of features, a target vector, an ML algorithm, a model performance parameter and transformation functions. Since the set of transformations require training and evaluation of all possible models for performance check, the method becomes computationally infeasible. Thus, different approaches have been adopted in tackling this issue.

The first approach was introduced by Kanter and Veeramachaneni [2015], which can be termed as expand-reduce method and helped in the development of data science machine. In this method, all the feature transformations are applied at one go to obtain all combinations of features, followed by feature selection and tuning of the hyper-parameters. Although there is only one modelling step apart from the feature selection, the method does not consider function compositions and a large number of features have to be considered in the feature selection step for evaluation of the performance. Several expand-reduce methods include AutoLearn [Kaul et al., 2017],

ExploreKit [Katz et al., 2016] and One Button Machine (OBM) [Lam et al., 2017]. With time, many other open-source platforms involving algorithms such as Feature tools and FeatureHub [Smith et al., 2017] were developed based on collaboration and contribution of skilled data scientists in providing codes for feature engineering.

The second approach is the genetic algorithm (evolutionary algorithmic technique) involving programmed elements known as 'chromosomes', which can improve performance and alter with time based on actual information from predefined tasks. A genetic programming based on tree-based representation was proposed by Tran et al. [2016] that can perform both feature selection and feature construction. This approach provided a slight enhancement in computational speed than the expand-reduce method.

Other methods for problems in feature engineering include meta-learning [Nargesian et al., 2017], reinforced learning [Khurana et al., 2018] and hierarchical organization of transformation [Khurana et al., 2016].

### 1.3.2 Hyper-Parameter Optimization

Hyper-parameters are one of the two types of parameters in an ML model which are set manually before actually training the model, contrary to the normal parameters that are optimized during the training. Hyper-parameters are generally model-specific and need to be set automatically in an AutoML platform for an optimized model performance, which makes it an important task [Hoos et al., 2014]. In order to make the technique understandable and accessible to non-technical individuals who lack expertise on ML applications but have the required domain knowledge, automatic hyper-parameter selection method has been proposed by computer researchers. Though the method brings out a better performance of the ML models with optimized and effective combinations of the hyper-parameters, it is bounded by limited usage of memory and search time. In order to tackle this issue, a number of hyper-parameter optimization methods have been adopted, as described below.

The first method is the grid search method, also termed as brute force method, which assumes no search space and the user specifies a set of values for individual hyper-parameters, thereby evaluating the set of Cartesian products. The suitability of this model is hampered by iterative increase of the configuration search space, which makes it a time-consuming method. Another alternative is the random search method where the sampling hyper-parameter configuration values are specified and set by users until the search budget is exhausted. Random search is computationally feasible, though it does not search all the configurations as in the case of grid search [Bergstra and Bengio, 2012].

Another method that can be termed as 'optimization from samples' for hyper-parameter optimization improves the new set of configuration spaces after every iteration considering the performance of the previous configurations. Two important methods in this type of method are the particle swarm optimization (PSO) and evolutionary optimization, both of which mimic the biological behaviour of humans (interacting socially and individually). PSO [Escalante et al., 2009] is based on updating the configuration space by moving towards the best individual configuration (solution) from the previous iterated solution and searching for neighbouring suitable configurations. On the other hand, evolutionary optimization [Back, 1996] manages a generation of population or configuration space having suitable individuals, improved over generations by introducing mutations (adopting the qualities of the previous individuals) and crossover (interchanging qualities between different individuals)into the system.

Bayesian optimization is another widely used recent method, which is suitable for application in AutoML frameworks and is an iterative and probabilistic algorithm type. It involves two components: surrogate model and acquisition function. The surrogate model is basically a probabilistic one employed for mapping the hyper-parameter configurations to their performances with some uncertainty. Acquisition function maintains a balance between exploration and exploitation within the search space of hyper-parameter configurations by determining the utility of the configurations. Bayesian optimization has been found to appropriately justify the hyper-parameter configuration performance and is widely in practice [Snoek et al., 2012; Dahl et al., 2013; Snoek et al., 2015; Klein et al., 2016; Melis et al., 2017]. Few open-source Bayesian optimization includes Hyperopt [Bergstra et al., 2013], Spearmint [HIPS/Spearmint, 2020], Talos [Autonomiotalos, 2020], Hyperas [maxpumperia/hyperas, 2020], SMAC [Hutter et al., 2011] and Google Vizier [Golovin et al., 2017].

### 1.3.3 Pipeline Optimizer

In the above-mentioned methods, only one component of the ML pipeline has been attempted to handle, while in a pipeline optimizer both the selection of the ML algorithm and the hyper-parameters are considered for a complete AutoML solution. Auto-WEKA [Thornton et al., 2013] is the first pipeline optimizer and an AutoML system which can simultaneously select ML algorithm among the different ML algorithms and optimize hyper-parameters. The creators have termed it as combined algorithm selection and hyper-parameter optimization (CASH) where selection of ML algorithms have also been assumed to be a hyper-parameter problem. CASH formulation facilitated the authors to adopt the Bayesian optimization methods, which gave high-quality results in a limited amount of time. The CASH problem can be solved employing SMAC [Hutter et al., 2011] optimization algorithms in an Auto-WEKA platform. Auto-WEKA has been upgraded to Auto-WEKA 2.0 [Kotthoff et al., 2017] by the creators with incorporation of all the improvements in the system which supports regression problem optimization, fully integrated into the WEKA system rather than being standalone software.

Auto sklearn [Feurer et al., 2015] is another pipeline optimizer for AutoML systems which is based on Scikit-Learn [Pedregosa et al., 2011], a python ML library, specializing in solving the CASH problem and incorporating improvements into the previous AutoML system. The improvements include meta-learning, based on efficient implementation of Bayesian optimization procedure and automated ensemble construction of optimized models. Auto sklearn can perform well in about 86% of cases than Auto-WEKA system and won the first place in the ChaLearn AutoML challenge [Guyon et al., 2015].

Tree-based pipeline optimization tool (TPOT) is another pipeline optimizer in an AutoML system based on open-source genetic programming and is employed for handling pre-processing of features, selection of models and optimization of hyper-parameters. Few of the ML operators included in TPOT system (a python ML library) include classification operators (logistic regression, random forest), feature pre-processing (polynomial featurization, scalers) and selection (RFE, variance thresholders, etc.) operators. These operators are treated as genetic programming primitives and combined for a full-fledged pipeline and from them genetic programming trees are constructed. In the python package DEEP [Fortin et al., 2012], the authors generated and optimized these pipelines using genetic programming algorithm, as described in Banzhaf et al. [1998]. TPOT system provides a flexible ML pipeline, and although the system is run by a random search mechanism, it provides a better accuracy and performance than by a guided search mechanism.

### 1.3.4 Neural Architecture Search (NAS)

With the recent developments in deep learning, ML has been able to solve more complex patterns in the field of healthcare systems by employing a neural network model. The model consists of multiple layers of nodes (or neurons), namely input, output and hidden layers, with no specific limitations for the number of hidden layers, due to which the entire model can be assumed to be a black box. Deep learning has been effectively employed in perceptual learning tasks, which include speech recognition [Hinton et al., 2012; Graves et al., 2013], visual recognition [Krizhevsky et al., 2012; Simonyan and Zisserman, 2014], language processing [Collobert et al., 2011; Bordes et al., 2014] and genomics (containing large volume of data) [Alipanahi et al., 2015; Asgari and Mofrad, 2015]. The input layer contains numerical representations of data, the output gives the prediction and the hidden layers carry out non-linear transformation of data. The networks can be highly complex in nature, with a large number of parameters to be trained and the performance dependent on the choice of hyper-parameters and the architecture of the network [Yue-Hei Ng et al., 2015; He et al., 2016]. With the widespread application of deep learning and its success in many fields, the neural architecture search has gained tremendous importance and is being adopted by AutoML researchers (data scientists) to manually design complex architecture. The objective is to search the NAS design which is the best for the corresponding set of training data and task. NAS can be well characterized by three factors – the search space, search strategy and performance estimation strategy. Searchspace is probably the best neural architecture which can be determined by NAS algorithms. Search strategy determines how the search spaces or the neural architectures are to be explored and is generally achieved by balancing trade-offs between exploration and exploitations [March, 1991]. Performance estimation strategy determines the performance of the NAS algorithms to find the most suitable neural architecture for a given training data set and by validating with the test data.

## 1.4 MACHINE LEARNING IN ENTERPRISE

With growing investments in cognitive and AI by enterprise and banking sectors (forecasted to be 77.4 billion dollars in 2022 as compared to 24 billion dollar in 2018 [IDC, 2018]), there is a vast requirement of data engineers and ML experts in the coming decade. Over half the numbers of data engineers in enterprises are engaged in the early stage of ML, with the rest of them actually involved in advanced stage and deployment of ML models to production [O'Reilly, 2018]. Applications of ML models in enterprises can be categorized mainly into three types – clustering, classification and prediction [Lee and Shin, 2019].

### 1.4.1 Clustering

Clustering basically means grouping of different sets of objects (that are not labelled initially) and is based on the similarities in multi-dimensional aspect of each group. The similarity of the objects can be evaluated and grouped by employing a similarity distance function, which are then analyzed for hidden patterns to be recognized, improving efficiency and time management. In businesses, clustering facilitates customer grouping based on their frequent choices and preferences, which provides a better recommendation for services and personalized products. Customer feedbacks and comments are also widely used data for clustering models for improved customer satisfaction. Netflix is a major user of clustering ML [Najafabadi et al., 2017] that enabled its 130 million global members to be clustered into 1,000 communities and send recommendations based on their taste for online content (TV shows and movies). Decision regarding storage and location facilities by the enterprises are also taken into account by employing clustering ML based on the purchasing behaviour of customers such as brand preferences and sales of similar products.

### 1.4.2 Classification

ML classifiers generally identify the class and categories of a new object or customer preference after training and testing the model extensively with already known classes and categories of objects. Binary classification and multiclass classification tasks are some of the applications of ML classifiers frequently used. Classification of a loan applicant to be credit trustworthy or not is a type of binary classification, while classification of customers' reviews to be categorized into high positive, positive, neutral, negative and high negative is a type of multiclass classification task. Thus, ML classifier enables retailers to analyse customer data efficiently and provide best services to them. Similarly, it enables banks to provide effective and rapid services to customers. ML classifiers are employed for classification of large volume of documents, as is done by State Street [State Street, 2018] using a combination of NLP algorithm and human editorial expertise in presenting a customized and ranked newsfeed to its clients by covering major global English news publications with data of investors' portfolios. ML classifiers have also been employed for optimizing business operations, such as a brewing company (Anheuser-Busch) which employs ML classifier to minimize cost of transportations, find the best route for the drivers [Chandran, 2018] and improve customer service by considering inputs from the drivers about the best time for delivery and parking availability. In retail industry, Walmart (U.S. retail store) employs ML classifier algorithm to detect the freshness of the vegetables and fruits by combining the role of product standards and specifications set by USDA and huge volume of raw photos, facilitating perishable goods to maintain its flow [Musani, 2018].

### 1.4.3 Prediction

Prediction is basically to identify patterns from the current large volume of data so that an accurate and most probable future event can be anticipated such as future market performance disruption. Prediction differs from the classification strategy in that in classification the current data is utilized in categorizing patterns and examining changes in them for any unusual events such as security breach or credit card fraud. ML predictions can also be employed for predictive maintenance of equipment and machines by learning the ML-generated data and developing predictive models over time. Similar is the case in GE motors where data about the behaviour of machine operation over time, anomalies, and trends in performances are identified and analyzed for required future maintenance [Forbes, 2017]. At Deere & Company [Grosch, 2018], ML predictions are employed for agricultural purposes to detect plants requiring pesticides and fertilizers and then providing them with adequate and optimal amount of those chemicals.

Since a single ML algorithm cannot give appropriate result for all random problems, the best one needs to be chosen based on the data variety, data speed, data volume and type of the problem itself, which can affect the performance of the algorithms. For the selection of the algorithm for a particular problem, trade-off between interpretability and accuracy must be the priority to manage. Accuracy generally represents how well the ML algorithms can identify the pattern and cause of the problem in terms of performance. Interpretability, on the other hand, is the ease with which a particular decision or response can be comprehended or which can be defined in terms of logical statements (such as if-then rules). The trade-offs between accuracy and interpretability can be realized due to involvement of two factors. One of the factors is that complex algorithms with numerous parameters are generally more accurate than the simpler ones, which make it difficult to understand and interpret by users. Another factor is that the highly accurate algorithms and models generally are devoid of any assumptions and are in the form of 'black box', making it impossible to understand how the problem gets solved and thus can be considered an unreliable outcome.

Apart from the data management challenges such as data sharing, accessibility, privacy and security, application of ML in enterprises faces some of the other higher level technical and management challenges, which includes ethical challenges, cost–benefit challenges, ML engineer shortage challenge and data quality challenges.

## 1.5 DISCUSSION

The ML algorithms in agriculture can be extended to distribution of quality products by predicting shipment delays, managing the distributor fleet, managing the vehicle delivery time and deciding vehicle pathways. Implementation of ML techniques in the processing phase of agricultural products can transform into data-driven and smart manufacturing systems with connected devices, advanced sensors, drones and intelligent systems. This will improve product traceability, transparency and supply chain visibility but will require the companies to heavily invest on ML technologies and practitioners to further explore such integration with blockchain and other technologies [Kamble et al., 2018; Sharma et al., 2018]. Few technical and non-technical challenges that have to be addressed by practitioners for successfully adopting ML in agriculture and for obtaining greater investments are data security, standardizing data, network infrastructure and interoperability within devices. Policymakers need to subsidize investments on ML technologies in agricultural sectors and provide farmers access to the latest technologies to make the implementation more affordable and robust.

The application of AutoML systems in healthcare sectors has been little due to complexity of ML algorithms, lack of transparency in blackboxes, assembling diverse and high-quality datasets, unreliability/variability of EHR datasets [Hersh et al., 2013; Elmore et al., 2017], partial interoperability of EHR systems [Polite et al., 2019] and inefficiency in ML pipeline optimization due to availability of large biomedical datasets. However, the introduction of interactive tool, as in ATM Seer systems by Wang et al. [2019], to better search, analyze and refine data spaces for AutoML systems can pave the way for newer path towards this implementation.

Although ML algorithms have been extensively employed in enterprises, a number of challenges in the domain of cost effectiveness, availability of ML experts and engineers, data refining, and ethics still persist. Enterprise managers must be accustomed to employing ML models in all real-world decision-making operations in the industry and must know the importance of refining the unstructured, multi-source information for high-quality data.

## 1.6 CONCLUSION

From the above discussion, we found out how ML algorithms can be adequately employed to agricultural, healthcare and enterprise sectors and what are the challenges faced during the implementation in the current scenario of technological revolutions, government policies, expertise in ML and data refining techniques. In agricultural sector, ML can be employed in various operational phases of an SASC such as pre-production, production, distribution and inventory. Numerous factors are associated with these phases which influence the outcomes of various decision-making processes that require different ML algorithms for better efficiency, accuracy and new pattern identification. The framework of ML implementation in agriculture has been discussed, including the adoption in three different components, i.e. the ASC, ML algorithms and SASC. In healthcare system, AutoML techniques can be employed to provide the practitioners the ability to carry out appropriate ML models for a specific decision problem involving a huge dataset of patient health conditions. AutoML techniques can be addressed based on the strategies adopted to automate problems in a healthcare setting, which are the automated feature engineering, hyper-parameter optimization, pipeline optimizer and neural architecture search. In enterprises, ML techniques have been widely employed with a higher number of data scientists actually involved in deploying this technology into operations. Here, the application of ML techniques can be mainly classified into threetypes, namely clustering, classification and prediction. Clustering technique basically involves unlabelled data for grouping of objects, while the classification involves labelled data for training the model first and then later using to classify new and unknown objects. Prediction involves all forms of data (labelled and unlabelled) to identify events that are most probable to occur in the future.

## REFERENCES

Adnan, N., Nordin, S. M., Rahman, I., & Noor, A. (2018). The effects of knowledge transfer on farmers decision making toward sustainable agriculture practices: in view of green fertilizer technology. World Journal of Science, Technology and Sustainable Development, 15(1), 98–115.

Ahn, H.S., Dayoub, F., Popovic, M., MacDonald, B., Siegwart, R., & Sa, I.(2018). An overview of perception methods for horticultural robots: from pollination to harvest. arXiv preprint arXiv:1807.03124.

Alipanahi, B., Delong, A., Weirauch, M. T., & Frey, B. J. (2015). Predicting the sequence specificities of DNA-and RNA-binding proteins by deep learning. Nature Biotechnology, 33(8): 831–838.

Ardila, D., et al. (2019). End-to-end lung cancer screening with three-dimensional deep learning on low-dose chest computed tomography. Nature Medicine, 25(5), 1.

Asgari, E., & Mofrad, M. R. (2015). Continuous distributed representation of biological sequences for deep proteomics and genomics. PLoS One, 10(11): e0141287.

Autonomiotalos (Retrieved on July 4, 2020). Available from: https://github.com/autonomio/talos

Back, T. (1996). Evolutionary algorithms in theory and practice: evolution strategies, evolutionary programming, genetic algorithms. New York, NY: Oxford University Press.

Banzhaf, W., Nordin, P., Keller, R. E., & Francone, F. D. (1998). Genetic programming: an introduction. Vol. 1. San Francisco, CA: Morgan Kaufmann.

Bates, D. W., Saria, S., Ohno-Machado, L., Shah, A., & Escobar, G. (2014). Big data in health care: using analytics to identify and manage high-risk and high-cost patients. Health Affair (Project Hope), 33(7):1123–1131.

Beam, A.L., & Kohane, I.S. (2018). Big data and machine learning in health care. JAMA, 319(13):1317–1318.

Bergstra, J., & Bengio, Y. (2012). Random search for hyper-parameter optimization. Journal of Machine Learning Research, 13: 281–305.

Bergstra, J., Yamins, D., & Cox, D. D. (2013). Hyperopt: a python library for optimizing the hyperparameters of machine learning algorithms. Proceedings of the 12th Python in Science Conference. DOI: 10.25080/Majora-8b375195-003.

Bordes, A., Chopra, S., & Weston, J. (2014). Question answering with subgraph embeddings. arXiv preprint arXiv:1406.3676.

Brown, N. (2015). Healthcare datagrowth: an exponential problem. Available from: https://www.nextech.com/blog/healthcare-data-growth-an-exponential-problem.

Buelvas Padilla, M.P., Nisperuza Canabal, P.A., López Pereira, J.M., & Hernández Riaño, H.E. (2018). Vehicle routing problem for the minimization of perishable food damage considering road conditions. Logistics Research, 11(2):1–18.

Chandran, P. (2018, July 23). Disruption in retail: AI, machine learning, and big data. Medium. Available from: https://towardsdatascience.com/disruption-in-retail-ai-machinelearning-big-data-7e9687f69b8f

Chlingaryan, A., Sukkarieh, S., & Whelan, B. (2018). Machine learning approaches for crop yield prediction and nitrogen status estimation in precision agriculture: a review. Computers and Electronics in Agriculture, 151:61–69.

Collobert, R., Weston, J., Bottou, L., Karlen, M., Kavukcuoglu, K., & Kuksa, P. (2011). Natural language processing (almost) from scratch. Journal of Machine Learning Research, 12:2493–2537.

Dahl, G. E., Sainath, T. N., & Hinton, G. E. (2013). Improving deep neural networks for LVCSR using rectified linear units and dropout. 2013 IEEE International Conference on Acoustics, Speech and Signal Processing. Vancouver, Canada, 26–31 May 2013.

Du, C.J., & Sun, D.W. (2006). Learning techniques used in computer vision for food quality evaluation: a review. Journal of Food Engineering, 72(1):39–55.

Elkington, J. (1998). Partnerships from cannibals with forks: the triple bottom line of 21st-century business. Environmental Quality Management, 8(1):37–51.

Elmore, J. G., et al. (2017). Pathologists' diagnosis of invasive melanoma and melanocytic proliferations: observer accuracy and reproducibility study. BMJ, 357: j2813.

Escalante, H. J., Montes, M., & Sucar, L. E. (2009). Particle swarm model selection. Journal of Machine Learning Research, 10: 405–440.

Escobar, G. J., et al. (2016). Piloting electronic medical record–based early detection of inpatient deterioration in community hospitals. Journal of Hospital Medicine, 11:S18–S24.

Feurer, M., Klein, A., Eggensperger, K., Springenberg, J., Blum, M., & Hutter, F. (2015). Efficient and robust automated machine learning. Advances in Neural Information Processing Systems, 2:2755–2763.

Forbes. (2017, June 7). How AI and machine learning are helping drive the GE digital transformation. Available from: https://www.forbes.com/sites/ciocentral/2017/06/07/how-ai-and-machine-learning-are-helping-drive-the-ge-digital-transformation/?sh=6e1d87a21686

Fortin, F.-A., De Rainville, F.-M., Gardner, M.-A., Parizeau, M., & Gagné, C. (2012). DEAP: evolutionary algorithms made easy. Journal of Machine Learning Research, 13(70):2171–2175.

Gharaei, A., Hoseini Shekarabi, S. A., & Karimi, M. (2019a). Modelling and optimal lot-sizing of the replenishments in constrained, multi-product and bi-objective EPQ models with defective products: generalised gross decomposition. International Journal of Systems Science: Operations & Logistics, 1–13. Available from: https://doi.org/10.1080/23302674.2019.1574364

Gharaei, A., Karimi, M., & Hoseini Shekarabi, S. A. (2019b). Joint economic lot-sizing in multiproduct multi-level integrated supply chains: generalized benders decomposition. International Journal of Systems Science: Operations & Logistics, 7(4):1–17.

Gharaei, A., Karimi, M., & Shekarabi, S. A. H. (2019c). An integrated multi-product, multi-buyer supply chain under penalty, green, and quality control polices and a vendor managed inventory with consignment stock agreement: the outer approximation with equality relaxation and augmented penalty algorithm. Applied Mathematical Modelling, 69, 223–254.

Golovin, D., Solnik, B., Moitra, S., Kochanski, G., Karro, J., & Sculley, D., (2017). Google Vizier: a service for black-box optimization. Proceedings of the 23rd ACM SIGKDD International Conference on Knowledge Discovery and Data Mining. DOI: 10.1145/3097983.3098043.

Graves, A., Mohamed, A.-r., & Hinton, G. (2013). Speech recognition with deep recurrent neural networks. 2013 IEEE International Conference on Acoustics, Speech and Signal Processing. Vancouver, Canada, 26–31 May 2013.

Grosch, K. (2018). John Deere: bringing AI to agriculture. Digital Initiative. Available from: https://digital.hbs.edu/platform-rctom/submission/john-deere-bringing-ai-to-agriculture/

Guyon, I., et al. (2015). Design of the 2015 ChaLearn AutoML challenge. 2015 International Joint Conference on Neural Networks (IJCNN). Killarney, Ireland, 12–17 July 2015.

Haghverdi, A., Washington-Allen, R.A., & Leib, B.G. (2018). Prediction of cotton lint yield from phenology of crop indices using artificial neural networks. Computers and Electronics in Agriculture, 152:186–197.

Hao, Y., Helo, P., & Shamsuzzoha, A. (2018). Virtual factory system design and implementation: integrated sustainable manufacturing. International Journal of Systems Science: Operations & Logistics, 5(2):116–132.

He, K., Zhang, X., Ren, S., & Sun, J. (2016). Identity mappings in deep residual networks. European Conference on Computer Vision. DOI: 10.1007/978-3-319-46493-0_38.

Hersh, W. R., et al. (2013). Caveats for the use of operational electronic health record data in comparative effectiveness research. Medical Care, 51(8 Suppl3):S30–S37.

Hinton, G., et al. (2012). Deep neural networks for acoustic modeling in speech recognition: the shared views of four research groups. IEEE Signal Processing Magazine, 29(6):82–97.

HIPS/Spearmint. (2020). (Retrieved on July 4, 2020). Available from: https://github.com/HIPS/Spearmint

Hutter, F., Hoos, H., & Leyton-Brown, K. (2014). An efficient approach for assessing hyperparameter importance. International Conference on Machine Learning. PMLR, 32(1):754–762.

Hoseini Shekarabi, S. A., Gharaei, A., & Karimi, M. (2019). Modelling and optimal lot-sizing of integrated multi-level multi-wholesaler supply chains under the shortage and limited warehouse space: generalised outer approximation. International Journal of Systems Science: Operations & Logistics, 6(3):237–257.

Hutter, F., Hoos, H. H., & Leyton-Brown, K. (2011). Sequential model-based optimization for general algorithm configuration. International Conference on Learning and Intelligent Optimization. Available from: https://ml.informatik.uni-freiburg.de/papers/11-LION5-SMAC.pdf

IDC. (2018). Worldwide spending on cognitive and artificial intelligence systems forecast to reach $77.6 billion in 2022, according to new IDC spending guide. Available from:

Im, J., Park, S., Rhee, J., Baik, J., & Choi, M. (2016). Downscaling of AMSR-E soil moisture with MODIS products using machine learning approaches. Environmental Earth Sciences, 75(15):1120

Kamble, S.S., Gunasekaran, A., & Gawankar, S.A. (2018). Sustainable Industry 4.0 framework: a systematic literature review identifying the current trends and future perspectives. Process Safety and Environmental Protection, 117:408–425.

Kamilaris, A., Kartakoullis, A., & Prenafeta-Boldú, F.X. (2017). A review on the practice of big data analysis in agriculture. Computers and Electronics in Agriculture, 143:23–37.

Kanter, J. M., & Veeramachaneni, K. (2015). Deep feature synthesis: towards automating data science endeavors. 2015 IEEE International Conference on Data Science and Advanced Analytics (DSAA). Paris, France, 19–21 October 2015.

Katz, G., Shin, E. C. R., & Song, D. (2016). Explorekit: automatic feature generation and selection. 2016 IEEE 16th International Conference on Data Mining (ICDM). Barcelona, Spain, 12–15 December 2016.

Kaul, A., Maheshwary, S., & Pudi, V. (2017). Autolearn—automated feature generation and selection. 2017 IEEE International Conference on Data Mining (ICDM). New Orleans, LA, 18–21 November 2017.

Kazemi, N., Abdul-Rashid, S. H., Ghazilla, R. A. R., Shekarian, E., & Zanoni, S. (2018). Economic order quantity models for items with imperfect quality and emission considerations. International Journal of Systems Science: Operations & Logistics, 5(2):99–115.

Khurana, U., Turaga, D., Samulowitz, H., & Parthasrathy, S. (2016). Cognito: automated feature engineering for supervised learning. 2016 IEEE 16th International Conference on Data Mining Workshops (ICDMW). Barcelona, Spain, 12–15 December 2016.

Khurana, U., Samulowitz, H., & Turaga, D. (2018). Feature engineering for predictive modeling using reinforcement learning. Thirty-Second AAAI Conference on Artificial Intelligence.arXiv:1709.07150.

Klein, A., Falkner, S., Bartels, S., Hennig, P., & Hutter, F. (2016). Fast Bayesian optimization of machine learning hyperparameters on large datasets. arXiv preprint arXiv:1605.07079.

Kotthoff, L., Thornton, C., Hoos, H. H., Hutter, F., & Leyton-Brown, K. (2017). Auto-WEKA 2.0: automatic model selection and hyperparameter optimization in WEKA. Journal of Machine Learning Research, 18(1): 826–830.

Krizhevsky, A., Sutskever, I., & Hinton, GE. (2012). ImageNet classification with deep convolutional neural networks. Advances in Neural Information Processing Systems, 25. DOI: 10.1145/3065386.

Kuo, C.C., et al. (2019). Automation of the kidney function prediction and classification through ultrasound-based kidney imaging using deep learning. NPJ Digital Medicine, 2(1):29.

Lam, H. T., et al. (2017). One Button Machine for automating feature engineering in relational databases. arXiv preprint arXiv:1706.00327.

Lee, I., & Shin, Y.J. (2019). Machine learning for enterprises: applications, algorithm selection, and challenges. Business Horizons, 63(2): 157–170.

Lee, J. H., You, J., Dobrota, E., & Skalnik, D. G. (2010). Identification and characterization of a novel human PP1 phosphatase complex. Journal of Biological Chemistry. Available from: http://www.jbc.org/cgi/doi/10.1074/jbc.M110.109801.

Liang, H., et al. (2019). Evaluation and accurate diagnoses of pediatric diseases using artificial intelligence. Nature Medicine, 25(3):433–438.

Lo-Ciganic, W.-H., et al. (2015). Using machine learning to examine medication adherence thresholds and risk of hospitalization. Medical Care, 53(8):720–728.

Lundberg, S. M., et al. (2018). Explainable machine-learning predictions for the prevention of hypoxaemia during surgery. Nature Biomedical Engineering, 2(10):749–760.

Luo, J., Wu, M., Gopukumar, D., & Zhao, Y. (2016). Big data application in biomedical research and health care: a literature review. Biomedical Informatics Insights, 8:1–10.

Maione, C., Batista, B.L., Campiglia, A.D., Barbosa Jr, F., & Barbosa, R.M. (2016). Classification of geographic origin of rice by data mining and inductively coupled plasma mass spectrometry. Computers and Electronics in Agriculture, 121:101–107.

Manzini, T., Lim, Y.C., Tsvetkov, Y., & Black, A.W. (2019). Black is to criminal as Caucasian is to police: detecting and removing multiclass bias in word embeddings. Proceedings of the 2019 Conference of the North American Chapter of the Association for Computational Linguistics. DOI: 10.18653/v1/N19-1062.

March, J. G. (1991). Exploration and exploitation in organizational learning. Organization Science, 2(1):71–87.

Marella, W. M., Sparnon, E., & Finley, E. (2017). Screening electronic health record–related patient safety reports using machine learning. Journal of Patient Safety, 13(1):31–36.

McNider, R.T., et al. (2014). An integrated crop and hydrologic modeling system to estimate hydrologic impacts of crop irrigation demands. Environmental Modelling & Software, 72:341–355.

Melis, G., Dyer, C., & Blunsom, P. (2017). On the state of the art of evaluation in neural language models. arXiv preprint arXiv:1707.05589.

Miotto, R., Li, L., Kidd, B. A., & Dudley, J. T. (2016). Deep patient: an unsupervised representation to predict the future of patients from the electronic health records. Scientific Reports, 6:26094.

Morellos, A., et al. (2016). Machine learning based prediction of soil total nitrogen, organic carbon and moisture content by using VIS-NIR spectroscopy. Biosystems Engineering, 152:104–116.

Murdoch, T.B., & Detsky, A.S. (2013). The inevitable application of big data to health care. JAMA, 309(13):1351–1352.

Musani, P. (1 March 2018). Eden: the tech that's bringing fresher groceries to you. Walmart. Available from: https://corporate.walmart.com/newsroom/community/20180301/eden-the-tech-thats-bringing-fresher-groceries-to-you

Najafabadi, M. K., Mahrin, M. N., Chuprat, S., & Sarkan, H. M. (2017). Improving the accuracy of collaborative filtering recommendations using clustering and association rules mining on implicit data. Computers in Human Behavior, 67:113e128.

Nargesian, F., Samulowitz, H., Khurana, U., Khalil, E. B., & Turaga, D. (2017). Learning feature engineering for classification. Proceedings of the Twenty-Sixth International Joint Conference on Artificial Intelligence (IJCAI-17).

O'Reilly. (2018). The state of machine learning adoption in the enterprise. Available from: https://www.oreilly.com/data/free/state-of-machine-learning-adoption-in-the-enterprise.csp

OECD. (2019). Technology and digital in agriculture. Available from: http://www.oecd.org/agriculture/topics/technology-and-digital-agriculture/

Özdemir, A., & Barshan, B. (2014). Detecting falls with wearable sensors using machine learning techniques. Sensors, 14(6):10691–10708.

Patidar, R., Agrawal, S., & Pratap, S. (2018). Development of novel strategies for designing sustainable Indian agri-fresh food supply chain. Sādhanā, 43(10):167.

Pedregosa F, et al. (2011) Scikit-learn: machine learning in Python. J Mach Learn Res, 12:2825–30.

Polite, B. N., et al. (2019). State of cancer care in America: reflections on an inaugural year. Journal of Oncology Practice, 15(4): 163–165

Porter, M.E., & Heppelmann, J.E. (2015). How smart, connected products are transforming companies. Harvard Business Review, 93(10):96–114.

Prasad, R., Deo, R.C., Li, Y., & Maraseni, T. (2018). Soil moisture forecasting by a hybrid machine learning technique: ELM integrated with ensemble empirical mode decomposition. Geoderma, 330:136–161.

maxpumperia/hyperas. (2020). (Retrieved on July 4, 2020) Available from: https://github.com/maxpumperla/hyperas.

Qiang, L., & Jiuping, X. (2008). A study on vehicle routing problem in the delivery of fresh agricultural products under random fuzzy environment. International Journal of Information and Management Sciences, 19(4):673–690.

Quanming Y, et al. (2018). Taking human out of learning applications: asurvey on automated machine learning. arXiv preprint arXiv:1810.13306.

Rabbani, M., Farshbaf-Geranmayeh, A., & Haghjoo, N. (2016). Vehicle routing problem with considering multi-middle depots for perishable food delivery. Uncertain Supply Chain Management, 4(3):171–182.

Rajkomar, A., Dean, J., & Kohane, I. (2019). Machine learning in medicine. New England Journal of Medicine, 380(14):1347–1358.

Rajkomar, A, et al. (2018). Scalable and accurate deep learning with electronic health records. NPJ Digital Medicine, 1(1):18.

Rajpurkar, P., et al. (2017). CheXNet: radiologist-level pneumonia detection on chest x-rays with deep learning. arXiv preprint arXiv:1711.05225.

Ribeiro, F.D.S., et al. (2018). An end-to-end deep neural architecture for optical character verification and recognition in retail food packaging. 25th IEEE International Conference on Image Processing (ICIP). Athens, Greece, 7–10 October 2018.

Rumsfeld, J. S., Joynt, K. E., & Maddox, T. M. (2016). Big data analytics to improve cardiovascular care: promise and challenges. Nature Reviews: Cardiology, 13(6):350–359.

Saria, S., Koller, D., &Penn, A. (2010). Learning individual and population level traits from clinical temporal data. Proceedings of Neural Information Processing Systems. Available from: https://citeseerx.ist.psu.edu/viewdoc/download?doi=10.1.1.232.390&rep=rep1&type=pdf

Sarkar, S., & Giri, B. C. (2018). Stochastic supply chain model with imperfect production and controllable defective rate. International Journal of Systems Science: Operations & Logistics, 7(2):133–146.

Sayyadi, R., & Awasthi, A. (2018a). A simulation-based optimisation approach for identifying key determinants for sustainable transportation planning. International Journal of Systems Science: Operations & Logistics, 5(2):161–174.

Shah, N. H., Chaudhari, U., & Cárdenas-Barrón, L. E. (2018). Integrating credit and replenishment policies for deteriorating items under quadratic demand in a three-echelon supply chain. International Journal of Systems Science: Operations & Logistics, 7(1):34–45.

Sharma, R., Kamble, S.S., & Gunasekaran, A. (2018). Big GIS analytics framework for agriculture supply chains: a literature review identifying the current trends and future perspectives. Computers and Electronics in Agriculture, 155:103–120.

Sharma, R., Kamble, S.S., Gunasekaran, A., Kumar, V., & Kumar, A. (2020). A systematic literature review on machine learning applications for sustainable agriculture supply chain performance. Computers and Operations Research, 119:104926.

Simonyan, K., & Zisserman, A. (2014). Very deep convolutional networks for large-scale image recognition. arXiv preprint arXiv:1409.1556.

Sirsat, M.S., Cernadas, E., Fernández-Delgado, M., & Barro, S. (2018). Automatic prediction of village-wise soil fertility for several nutrients in India using a wide range of regression methods. Computers and Electronics in Agriculture, 154:120–133.

Smith, M. J., Wedge, R., & Veeramachaneni, K. (2017). FeatureHub: towards collaborative data science. 2017 IEEE International Conference on Data Science and Advanced Analytics (DSAA). Tokyo, Japan, 19–21 October 2017.

Snoek, J., et al. (2015). Scalable Bayesian optimization using deep neural networks. International Conference on Machine Learning. Available from: https://arxiv.org/pdf/1502.05700.pdf

Snoek, J., Larochelle, H., & Adams, R. P. (2012). Practical Bayesian optimization of machine learning algorithms. Advances in Neural Information Processing Systems 25.arXiv:1206.2944.

State Street. (2018). State Street Verus: Frequently Asked Questions. Available from: http://www.statestreet.com/content/dam/statestreet/documents/SSGX/18-33087_Verus_FAQ.pdf

Tang, J., Wang, D., Zhang, Z., He, L., Jing, X., & Xu, Y. (2017). (2017). Weed identification based on K-means feature learning combined with convolutional neural network. Computers and Electronics in Agriculture, 135:63–70.

Thornton, C., Hutter, F., Hoos, H. H., & Leyton-Brown, K. (2013). Auto-WEKA: combined selection and hyperparameter optimization of classification algorithms. Proceedings of the 19th ACM SIGKDD International Conference on Knowledge Discovery and Data Mining. DOI: 10.1145/2487575.2487629.

Toga, A. W., et al. (2015). Big biomedical data as the key resource for discovery science. Journal of the American Medical Informatics Association, 22 (6):1126–1131.

Tran, B., Xue, B., & Zhang, M. (2016). Genetic programming for feature construction and selection in classification on high-dimensional data. Memetic Computing, 8(1):3–15.

Traore, B., Descheemaeker, K., Van Wijk, M., Corbeels, M., Supit, I., & Giller, K. (2016). Use of crop modelling to assess climate risk management for family food self-sufficiency in southern Mali. International Crop Modelling Symposium, Berlin, Germany, 15–17 March.

Wang, Q., et al. (2019). ATMSeer: increasing transparency and controllability in automated machine learning. arXiv preprint arXiv:1902.05009.

Wang, X.P., Wang, M., Ruan, J.H., & Li, Y. (2018). Multi-objective optimization for delivering perishable products with mixed time windows. Advances in Production Engineering & Management, 13(3):321–332.

Waring, J., Lindvall, C., & Umeton, R. (2020). Automated machine learning: review of the state-of-the-art and opportunities for healthcare. Artificial Intelligence in Medicine, 104:101822.

Yue-Hei Ng, J., et al. (2015). Beyond short snippets: deep networks for video classification. Proceedings of the IEEE Conference on Computer Vision and Pattern Recognition. Boston, MA, 7–12 June 2015.

Zhang, J., Song, F., & Tang, J. (2014). Computer modelling and new technologies, 18(10): 203–206.

# Section II

## Application towards Mechanical Engineering

# 2 Artificial Intelligence in Predicting the Optimized Wear Behaviour Parameters of Sintered Titanium Grade 5 Reinforced with Nano $B_4C$ Particles

*T. Ramkumar*
Dr. Mahalingam College of Engineering
and Technology Pollachi, India

*M. Selvakumar*
Dr. Mahalingam College of Engineering
and Technology Pollachi, India

**CONTENTS**



## 2.1 INTRODUCTION

Titanium and titanium alloys have several applications, such as in medical industries, automobile and aerospace industries. However, titanium has a very good corrosion and biocompatibility properties. Specific titanium grade 5 (Ti-6Al-4V) alloys have very unique properties, such as high strength and hardness [1-2]. To fabricate the composites, many techniques exist, such as liquid metallurgy, solid metallurgy and infiltration technique. Among these, solid metallurgy is used quite often in order to fabricate the composites because of low cost and ease of fabricating the net shape [3-4].

We study the effect of ultra-fine $B_4C$ powder on tribological behaviour of Ti-Al-V-x$B_4C$ at elevated temperature. It was observed that ultra-fine $B_4C$ particles dispersed evenly, resulting in a Ti-Al-V-$B_4C$ composite with tremendous increase in strength and tribological properties [5-7].

In this research, nano $B_4C$ was reinforced with titanium grade 5 (Ti-6Al-4V) in order to evaluate the tribological behaviour of the composite at elevated temperature conditions. Furthermore, the statistical analysis was conducted using full factorial design. In this design experiment, wt. % of $B_4C$, sliding distances, load and pin temperature were taken as factors, and specific wear rate (SWR) and coefficient of friction (CoF) were taken as responses.

## 2.2 MATERIALS AND METHODS

The precursors, viz. Ti, Al and V, having particle size of 44 μm and $B_4C$ having particle size of 10 μm were taken. The initial SEM morphology of the powders is shown in Figure 2.1(a–d). The figure shows that all particle sizes are in sub-micron level. Afterwards, the powders' sizes were reduced using ball milling machine. All the processes were carried out at inert atmosphere; also, toluene ($C_6H_5$–$CH_3$) was filled in the chamber in order to avoid oxidation. For every 0.25 hours, the chamber was checked and toluene was refilled to avoid oxidation. Ti, Al, V and $B_4C$ milled powder SEM images are displayed in Figure 2.1(g–h). However, to fabricate the composite samples, the composite powders were prepared by taking various wt. %, such as Ti-6Al-4V and Ti-6Al-4V-(2-10) $B_4C$, and the corresponding SEM images are shown in Figure 2.1(i–l).

### 2.2.1 Composite Preparation

The desired wt. % of Ti, A, V and $B_4C$ particles were kept in the ball mill chamber. The ball mill was rotated at 300 rpm to homogenously mix all the powder particles. After mixing, the composite powders were compacted. The die was coated with $MoS_2$ for lubricating purpose and easily ejecting the specimen. To make the green compacts, the composite powders were compacted using the compression testing machine. Later, the compacts were sintered using muffle furnace at the temperature of ± 900°C.

### 2.2.2 Elevated Temperature Wear Test

Using pinon disc apparatus, the elevated temperature wear test experiments were performed. Before conducting the experiments, the sample surfaces were polished

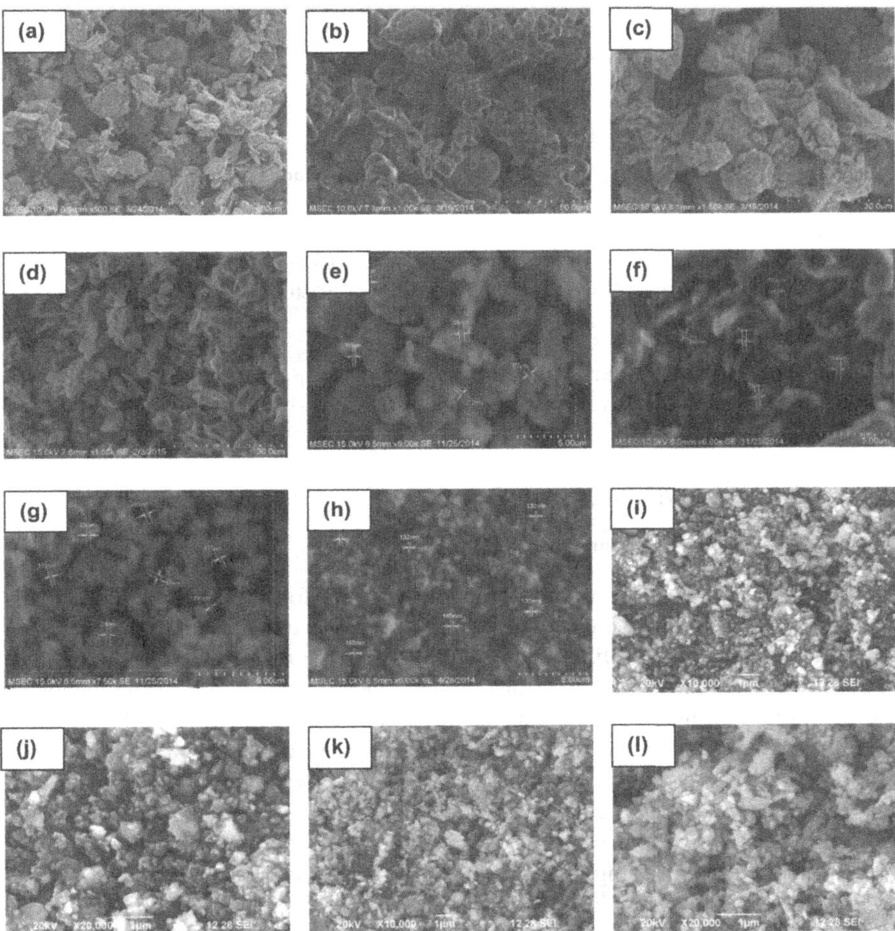


**FIGURE 2.1** SEM micrographs. (a–d) Ti, Al, V and $B_4C$ before milling. (e–h) Ti, Al, V and $B_4C$ after milling. (i–l) Various wt. % of $B_4C$ homogenously mixed with Ti-6Al-4V.

using emery paper in order to avoid the burs [8]. The sample weight was measured before and after conducting the experiment. Using the weight differences, the SWR was calculated. The CoF was calculated based on frictional force observed during the experimentation.

## 2.3 FULL FACTORIAL DESIGN

Full factorial design is a great tool that follows a technical methodology, which has been beneficial for design optimization. In this design, orthogonal array is used to evaluate the parameter and to reduce the number of experiments which were selected.

**TABLE 2.1**
**Data and its levels**

| | Factors | | | |
|---|---|---|---|---|
| **Level** | **Wt. % of $B_4C$** | **Applied load (N)** | **Sliding distance (m)** | **Pin temperature (°C)** |
| 1 | 0 | 10 | 1,000 | 30 |
| 2 | 4 | 15 | 1,400 | 50 |
| 3 | 8 | 20 | 1,800 | 100 |
| 4 | 10 | 25 | 2,200 | 150 |

For this study, the factors are sliding distance, pin temperature and load. Table 2.1 shows the factors and its levels selected for this study.

## 2.4 RESULTS AND DISCUSSION

### 2.4.1 Effect of Coefficient of Friction

Figure 2.2(a) indicates the normal probability graph for Ti-6Al-4V-$B_4C$ composite. The plot reveals that the model is significant because the data points are not deviated from the straight line. Figure 2.2(a) shows the residuals plot for CoF. The graph illustrates that the scatter points are very close to the zero line and the errors are negligible. Almost bell-shaped, symmetrical histogram can be inferred from Figure 2.2(a). The residual for the experimental measurements plot is shown in Figure 2.2(a). The plot illustrates that the order of observation has taken place as a line sequence in order to get the data The corresponding interaction plot is shown in Figure 2.2(b) for various process parameters for the occurrence of CoF. The wt. % of $B_4C$, applied load, sliding distance and specimen temperature were fixed at level 4. However, the optimum CoF was observed for higher amount of $B_4C$ addition, as confirmed from scatter plot [9].

The contour plot for CoF is presented in Figure 2.3(a,b) – Figure 2.3(a) shows load vs wt. % of $B_4C$ and Figure 2.3(b) shows S.D. vs wt. % of $B_4C$. From the plot, it is clear that it is a function of three factors, namely boron carbide wt. %, pin temperature and sliding distance. Keeping the applied load as constant. Figure 2.3(a) shows the contour plot for CoF at 25-N loads. The green colour was appearing as leading one for CoF. The blue region occurs for 20-N load and shows that the CoF amplified with the rise in sliding distance. From Figure 2.3(a, b), green and blue distinct zones were observed for the wear map at 20-N load. It illustrates that the decrease in SWR for increasing amount of $B_4C$ [10].

### 2.4.2 Effect of Specific Wear Rate

Figure 2.4(a) indicates the normal probability graph for Ti-6Al-4V-$B_4C$ composite. The plot reveals that the model is significant because the data points are not deviated from the straight line. Figure 2.4(a) shows the residual plot for SWR. The

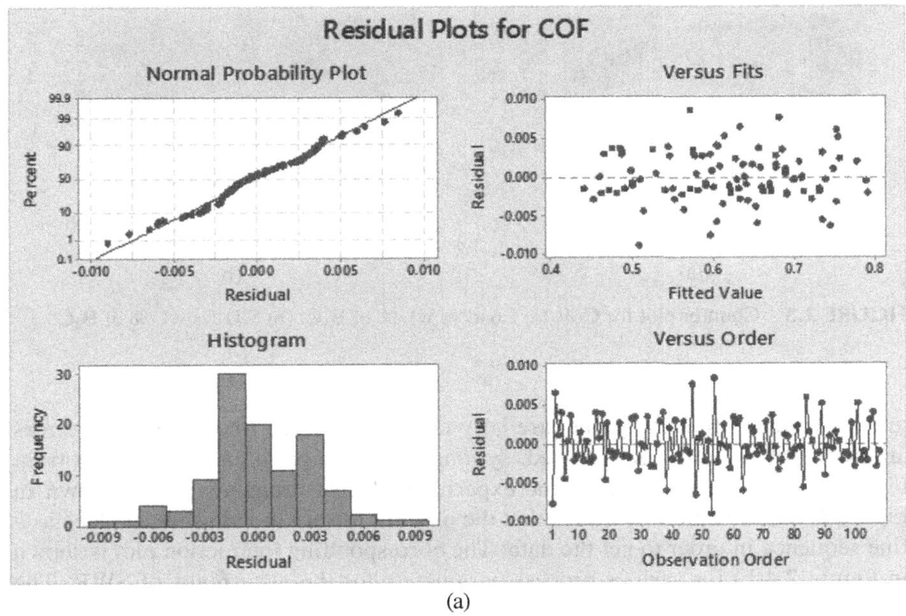


(a)

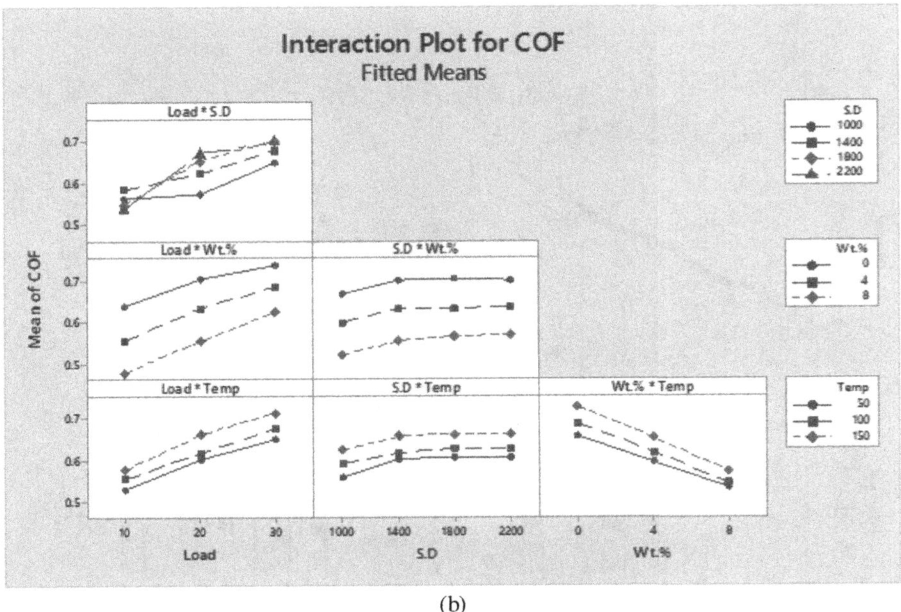


(b)

**FIGURE 2.2** (a) CoF residual plot. (b) CoF interaction plot.

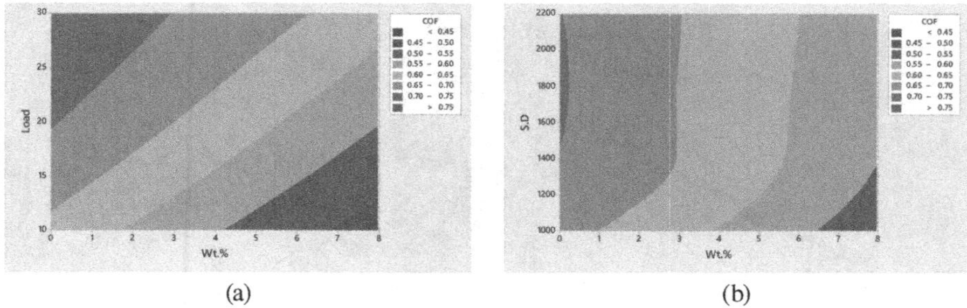


(a) (b)

**FIGURE 2.3** Counter plot for CoF. (a) Load vs wt. % of $B_4C$. (b) S.D. vs wt. % of $B_4C$.

graph illustrates that the points are very close to the zero line; hence, the errors are negligible. Almost bell-shaped, symmetrical histogram can be inferred from Figure 2.4(a). The residual for the experimental measurements plot is shown in Figure 2.4(a). The plot illustrates that the order of observation has taken place as a line sequence in order to get the data. The corresponding interaction plot is shown in Figure 2.4(b) for various process parameters for the occurrence of SWR. The wt. % of $B_4C$, applied load, sliding distance and specimen temperature are fixed at level 4. However, the optimum SWR was observed for higher amount of $B_4C$ addition, as confirmed by scatter plot [9].

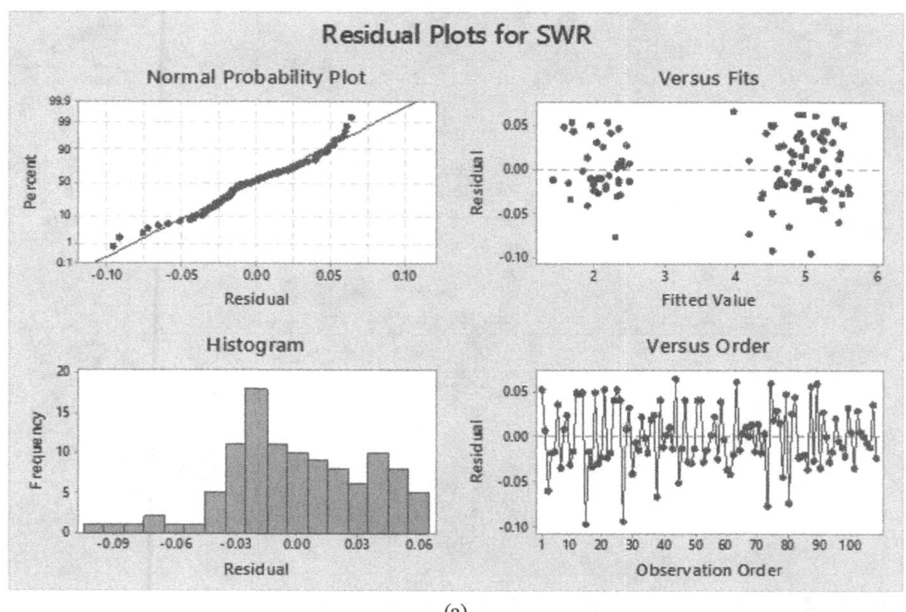


(a)

**FIGURE 2.4** (a) SWR residual plot. (b) SWR interaction plot.

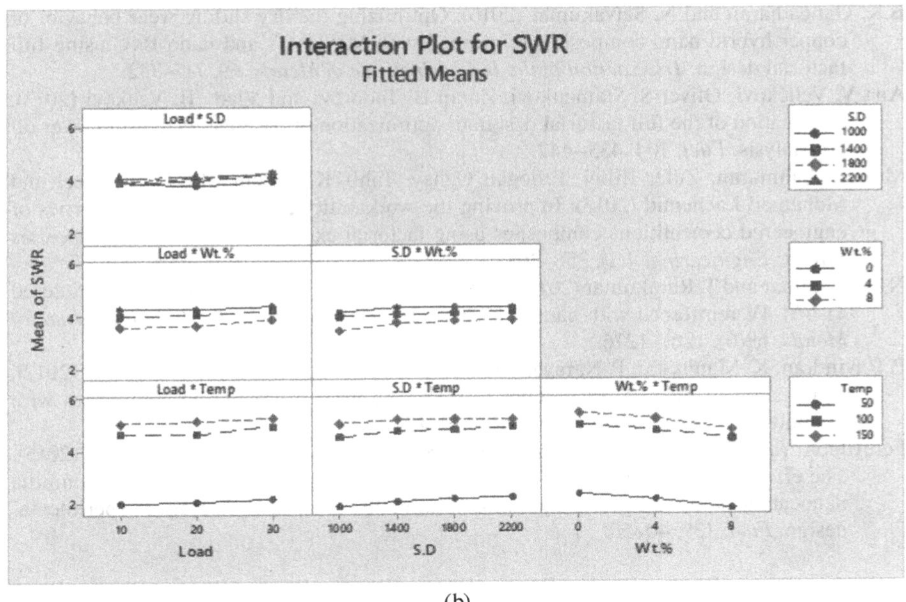


(b)

**FIGURE 2.4** *(Continued)*

## 2.5 CONCLUSION

- SEM micrographs show the ultra-fine $B_4C$ particles have homogenous distribution in the matrix material.
- The SWR and CoF of Ti-6Al-4V-$B_4C$ composites are reduced by adding 10 wt. % of $B_4C$.
- The applied load, sliding distance and wt. % of ultra-fine $B_4C$ particles have very less effect on SWR, but the load applied and pin temperature more expressively upset the CoF.

## REFERENCES

Heeman Choe, Susan M. Abkowitz, Stanley Abkowitz and David C. Dunand (2005), Effect of tungsten additions on the mechanical properties of Ti-6Al-4V, *Materials Science and Engineering,* A 396, 99–106.

T.M.T Godfrey, A. Wisbey, P.S. Goodwin, K. Bagnall and C.M. Ward-Close (2000), Microstructure and tensile properties of mechanically alloyed Ti-6Al-4V with boron additions, *Materials Science and Engineering,* A 282, 240–250.

S.C. Vettivel, N. Selvakumar, R. Narayanasamy and N. Leema (2013), Numerical modelling, prediction of Cu–W nano powder composite in dry sliding wear condition using response surface methodology, *Materials & Design,* 50, 977–996.

A.P. Ravindran, K. Manisekar, P. Rathika and P. Narayanasamy (2013), Tribological properties of powder metallurgy – processed aluminium self-lubricating hybrid composites with SiC additions, *Materials & Design,* 45, 561–570.

B.K. Gangatharan and N. Selvakumar (2016), Optimizing the dry sliding wear behavior of copper hybrid nano composites reinforced with MWCNTs and nano B4C using full factorial design, *Transaction of the Indian Institute of Metals*, 69, 717–732.

Ana V. Veličkovi, Oliver S. Stamenkovi, Zoran B. Todorovi and Vlada B. Veljkovi (2013), Application of the full factorial design to optimization of base-catalyzed sunflower oil ethanolysis, *Fuel,* 104, 433–442.

Mustafa Şahmaran, Zafer Bilici, Erdogan Ozbay, Tahir K. Erdem, Hasan E. Yucel and Mohamed Lachemid (2013), Improving the workability and rheological properties of engineered cementitious composites using factorial experimental design, *Composites Part B: Engineering,* 4(1), 356–368.

N. Selvakumar and T. Ramkumar (2016), Effect of high temperature wear behavior of sintered Ti-6Al-4V reinforced with nano B4C particles, *Transaction of the Indian Institute of Metals,* 69(6), 1267–1276.

P. Ravindran, K. Manisekar, P. Narayanasamy, N. Selvakumar and R. Narayanasamy (2012), Application of factorial techniques to study the wear of Al hybrid composites with graphite addition, *Materials & Design,* 39, 42–54.

Fereydoon Yaripour, Zahra Shariatinia, Saeed Sahebdelfar and Akbar Irandoukht (2015), The effects of synthesis operation conditions on the properties of modified γ-alumina nanocatalysts in methanol dehydration to dimethyl ether using factorial experimental design, *Fuel,* 139, 40–50.

# 3 Tribological Behaviour of AL7068-Alumina-B4C Hybrid Composites and Optimization with DEMATEL Technique

*Mohan Kumar Pradhan*
Maulana Azad National Institute of Technology
Bhopal, India

*Md. Samar Waheed*
Maulana Azad National Institute of Technology
Bhopal, India

*Shubham Gupta*
Maulana Azad National Institute of Technology
Bhopal, India

**CONTENTS**

## 3.1 INTRODUCTION

Traditional monolithic products have constraints on ensuring a successful blend of mechanical properties, along with stiffness, hardness and density. Composite materials can help in reducing these shortcomings and in meeting the increasing demands of modern technology. Metal matrix composites (MMCs) have considerably improved features, in addition to having high specific strength, compared with non-reinforced alloys. They have a higher specific modulus, a damping capability and a decent resistance to wear. There is a growing concern regarding composite materials with reduced density and less cost reinforcement materials.

Today, particle-reinforced aluminium-based composites are becoming increasingly important due to their low cost, isotropic advantages and the possibility of promoting secondary component manufacturing through secondary processing. Compared with the unreinforced alloy, the cast aluminium-based particle-reinforced composite material has a higher strength-to-weight ratio, specific stiffness and great resistance to wear. Low-density fly ash may help produce lightweight insulating composite materials. The granular composite material might well be processed by inserting the reinforcement particles into liquid matrix by casting into a molten metallurgical path. The casting route is favoured, as it is economical and ideal for large-scale manufacturing. Over the whole line of liquid production, stirring casting is the easiest and of lowest cost. The exclusive concern cognate to the process is the unequal arrangement of particulate matter owing to reduced wetting efficiency and the controlled division of gravity. The mechanical characteristics of the reinforced material are influenced by the size, form and composition of the structural material, as well as the surface structure and the interfacial reactions.

## 3.2 COMPOSITE MATERIALS

A blend of two or even more entities assembled on a microscopic level is a composite textile. A composite material consists of fabric, stones, fillers and/or particles incorporated into the structure, such as polymers, ceramics and metals. The matrix incorporates the refurbishment to optimize the necessary material structure, whilst the refurbishment increases the mechanical and thermal characteristics. The latest hybrid material provides greater strength than a single material when built properly. Various materials, along with metals, doped ceramics, polymers and chemical mixtures, have phases of their composition distributed in limited amounts. These are, however, not regarded as a composite element, as their features remain the same as their main components. The desired properties of the composite material are high tensile strength, stiffness, low density, good electrical conductivity but also thermal conductivity, reasonable friction coefficient, elevated temperature stability, corrosion and wear resistance, etc. Composite materials are multifunctional and provide properties and characteristics that cannot be derived from any single material. The materials used to make composite materials should be compatible even if they differ in composition and form.

## 3.3 CLASSIFICATION OF COMPOSITE

Composite products can be categorized in two distinct ways:

1. Based on the matrix material.
2. Based on the form of reinforcement.

### 3.3.1 Based on Matrix Material

#### 3.3.1.1 Metal Matrix Composite

MMCs are composed of at least two chemically and mechanically distinct layers, together with all composite products dispersed uniformly in such a way that properties cannot be obtained from any of them. MMC is a combination of two components, i.e. a metal base with a reinforcement component, which is generally non-metallic and is often ceramic. By definition, the manufacturing process for MMCs is different from conventional metal alloys. Compared to their polymer matrix counterparts, MMCs are often manufactured by combining two components (e.g. metal and ceramic fibres). Specific forms of MMCs are

1. Aluminium matrix composites (AMC).
2. Magnesium matrix composite.

3. Titanium matrix composite.
4. Copper matrix composites.

Among many metallic alloys, aluminium alloys are the most widely used as a matrix for the MMCs. The Al alloys are widely used owing to their desired properties, such as low density, strong corrosion strength, excellent electricals as well as thermal conductivity, and potential to increase precipitation, including their high damping efficiency. AMCs have been extensively researched, and are now being utilized in various applications. Sports products, fittings, packaging of electronic products, automotive components, etc., offer a wide range of properties, particularly mechanical, obtained from different matrix compositions – generally reinforced with alumina ($Al_2O_3$), silicon carbide (SiC), and carbon. The following methods are used for the manufacturing of the aluminium metal matrix.

1. Powder metallurgy.
2. Stir casting.
3. Squeeze casting.

#### 3.3.1.2 Ceramic Matrix Composite

Ceramic matrix composite (CMC) consists of ceramic particles reinforced by certain ceramic material fibres in the matrix just at the dispersed stage. CMCs have high toughness than the traditional ceramics, as traditional ceramics are highly brittle. CMC is usually reinforced by either continuous or long fibres or even by discontinuous or short fibres.

Ceramic composites reinforced by short discontinuous fibre are commonly made using traditional methods such as aluminium oxide or non-oxide silicon carbide ceramic matrix reinforced by SiC, aluminium nitride (AlN), zirconium oxide ($ZrO_2$), titanium boride ($TiB_2$), and other ceramic fibres. Many ceramic metal composites are typically filled with SiC fibres since they have higher strength as well as stiffness. Whiskers embedded in a short-fibre form improve the toughness and resistance to cracks propagation in CMC. However, short-fibre-reinforced fibre is prone to failure.

Long-fibre-reinforced ceramic composites are embedded with long monofilament or long multi-filament fibres. Continuous monofilament fibre provides higher strengthening effect in dispersed form. These are made by chemical vapour deposition of SiC on a base composed of tungsten or carbon fibre. Monofilament fibres have a good interfacial attachment to the matrix material that increases toughness.

#### 3.3.1.3 Polymer Matrix Composite

Polymer matrix composites (PMCs) are materials made by mixing a polymer (resin) matrix with a fibre-reinforced dispersion. Because of their inexpensiveness and simplified approaches to manufacturing, polymer-based composites are fairly popular. The usage of unreinforced polymers for structural materials is constrained by their poor mechanical properties as well as by their poor strength (the tensile strength of a few of the strongest polymers epoxy resins is 140 MPa). In contrast to its relatively poor strength, polymer products do indeed have poor impact resistance.

### 3.3.2 Based on Form of Reinforcement

#### 3.3.2.1 Particulate Composite

There are two types:

1. Particulate composites with randomly oriented particles.
2. Composite material with the preferential orientation of the particle. The dispersed phase of such materials comprises two-dimensional flat flakes parallel to each other.

The impact of dispersed particles on its composite properties relies upon the scale of the particle. Tiny fragments (less than 0.25 microns in diameter) were precisely dispersed in the matrix, avoiding dislocation and deformation of the substance. The reinforcing mechanism is just like the precipitation hardening effect. In precipitation hardening, as precipitation particles melt in the matrix precipitation hardening occurs at elevated temperatures, whereas the scattered process of granular composites becomes typically robust at high temperatures, maintaining the reinforcement effect. Some composite materials have been developed to be applied in high-temperature applications. Widely distributed phase particles have a lower binding effect but are capable of spreading the load uniformly to the substrate, resulting in improved stiffness and decreased ductility. Hard particulates scattered in a softer matrix can improve resistance to attrition. Softly scattered particulates in a stronger substrate increase their workability (leading fragments in a steel or copper substrate) and friction coefficient. The particle reinforcement and flake reinforcement are depicted in Figures 3.1 and 3.2, respectively.

#### 3.3.2.2 Fibrous Composite

Short-fibre-reinforced composites are made of a distributed phase reinforcement matrix in the shape of discrete fibres.

1. Composite with randomly oriented fibres.
2. Composite material having a favoured orientation of the fibres.

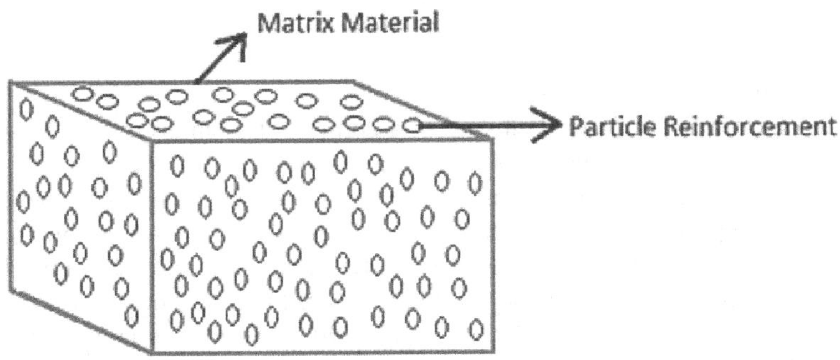


**FIGURE 3.1** Particle reinforcement.

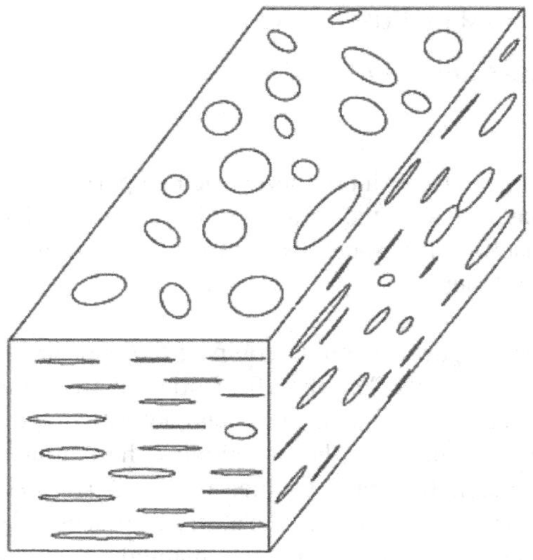

**FIGURE 3.2** Flake reinforcement.

Long-fibre-reinforced composites comprise of a matrix in which the dispersed phase is reinforced with continuous fibre.

1. The unidirectional orientation of the fibres.
2. The bidirectional orientation of the fibres.

If the fibers are oriented in a certain direction, stress is induced in the same direction, the impact of increasing strength becomes more pronounced. The reinforcing effect of long-fibre-reinforced composite is higher than that of the short-fibre-reinforced composite. The short-fibre-reinforced composites comprise of a matrix improved in the form of fibres with a separated process and have limited ability to share loads. The load provided mostly to the lengthy fibre-reinforced composite component is carried mainly by the scattered phase fibre. The matrix in this material acts almost like a fibre-binding material, holding it in the proper form as well as shielding it from mechanical or chemical damage.

#### 3.3.2.3 Laminate Composite

If a fibre-reinforced composite substance is composed of many layers of specific fibre orientation, it is referred to as a multilayer composite. Laminated composite materials have improved mechanical strength in two directions at perpendicular mostly to the required direction of the fiber or sheet and in one direction only, and the material has poor mechanical properties. Laminated composite materials consist of two-dimensional sheets/panels with the desired high-strength path, including wood or even continuous and balanced reinforced fibre plastics (Figure 3.3). The layers are layered and then blended in a way that each subsequent layer shifts the inclination of the high-strength surface.

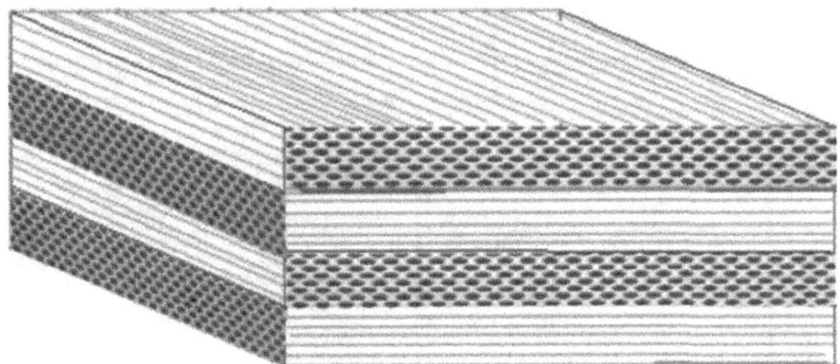

**FIGURE 3.3** Laminated composite.

## 3.4 PROCESSING OF COMPOSITES

MMCs may be fabricated through many technologies. Picking out the right process engineering focuses on the required types, quantities and distributions of reinforcing components (particles and fibres), matrix alloys and applications. By changing the fabrication method, processing and finishing, and by reinforcing the form of the component, different characteristic curves can be obtained despite involving the same piece and components. MMCs can be built up by a liquid- or solid-state process.

### 3.4.1 Liquid State Fabrication of Metal Matrix Composites

Liquid preparation of metal-based composites includes incorporating the scattered process into the melted base metal and then solidifying it. Effective interfacial adhesion (wetting) between the scattered process and the liquid matrix will be achieved to support the strong mechanical properties of the composite. Increased wettability could be obtained by covering particles (fibres) with scattered chemicals. The appropriate coating may not only reduce the amount of energy from the interfaces but also preclude the chemical reaction between the dispersed phase and the matrix.

The techniques used to render MMCs liquid state are:

1. Stir casting.
2. Infiltration.
3. Gas pressure infiltration.
4. Squeeze casting infiltration.
5. Pressure die infiltration.

#### 3.4.1.1 Stir Casting

This is a liquid-state process for the manufacture of composite materials where the scattered layer (ceramic fragments, thin fibres) is combined with molten metal matrix through mechanical stirring. Stir casting is the easiest or even often cost-effective process for processing liquids. Stir casting set-up, as depicted in Figure 3.4, consists of a furnace, a reinforcement feeder and a mechanical stirrer. The liquid composite

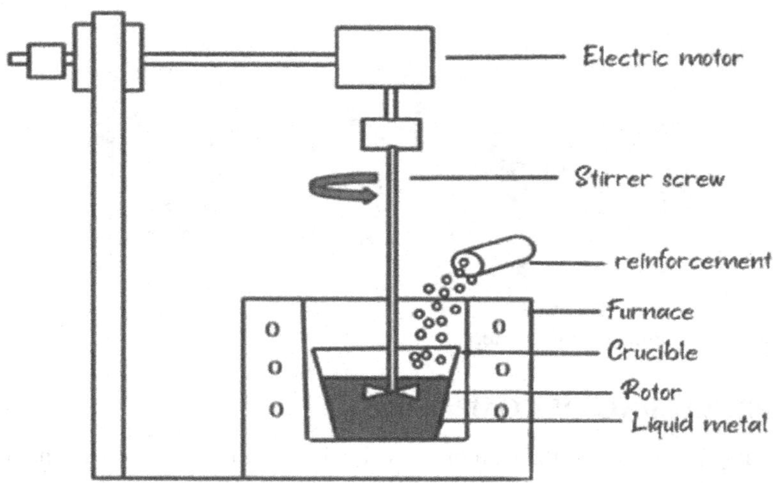


**FIGURE 3.4** Stir casting machine (Kumar and Pradhan, 2016).

material is cast using traditional casting processes and may also be handled via modern metal-shaping techniques. In this method, particles begin to form agglomerates that could be separated through intense stirring. However, here it is necessary to prevent gas from entering the melt, as this may lead to undesirable porosity or reactions. Extra precautions need to be taken for improving the dispersibility of the components in such a way that perhaps the susceptibility of the elements is matched with both the temperature of the melt and the length of the stirring, since this reaction with the melt will contribute to the breakdown of the reinforcement components. According to the small surface area to the volume ratio of the spherical particles, the decreased susceptibility of the irritating particles is usually less important than the fibres.

#### 3.4.1.2 Squeeze Casting Infiltration

Infiltration extrusion casting method is indeed a forced penetration technique for the determination of the composite matrix in the liquid phase. A mobile part of the mould is used to apply pressure to the molten metal and to push it into the dispersed phase. The mould is placed in the lower part of the fixed mould. The press casting penetration process is close to the press casting methodology used in the construction of metal alloys. Squeeze casting or pressure casting has been the most general method of producing MMCs. By gradually entering the mould, the melt solidifies under very high pressure, forming a fine-grained structure. Compared to this method, in die-casting technique there is no gas that can heat the finished parts. Figure 3.5 depicts the detailed set-up of the squeeze casting infiltration.

#### 3.4.1.3 Pressure Die Infiltration

Die infiltration is a forced infiltration method of the metal die casting process. The die-casting technique is used to place the pre-formed dispersed phase (particles,

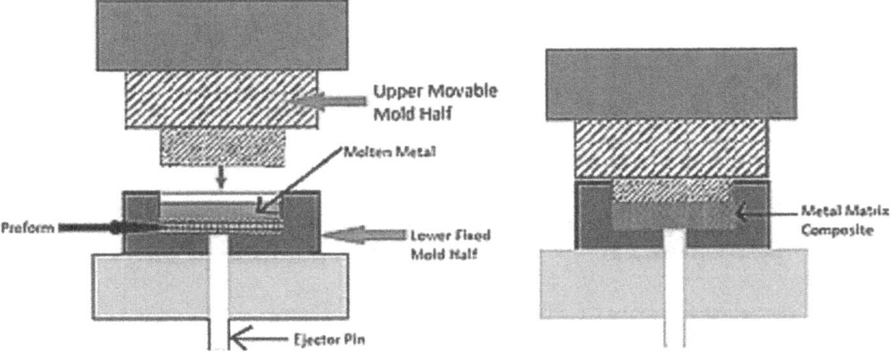


**FIGURE 3.5** Squeeze casting infiltration.

fibres) into the mould, and then the molten metal is filled into the mould. The mould passes through the gate and the pressure of the movable piston penetrates into the pre-form, as shown in Figure 3.6.

#### 3.4.1.4 Gas Pressure Infiltration

Air pressure infiltration is a forced infiltration process for processing MMCs during the liquid phase. It uses pressurized gas to exert pressure on the molten metal and forces it to enter the pre-formed dispersed process. In the case of gas pressure penetration, the melt reaches the pre-form by gas added from the outside. Using a gas that is volatile with regard to the ground. The melting and infiltration of the matrix are performed in an appropriate pressure vessel. The detailed schematic diagram is depicted in Figure 3.7.

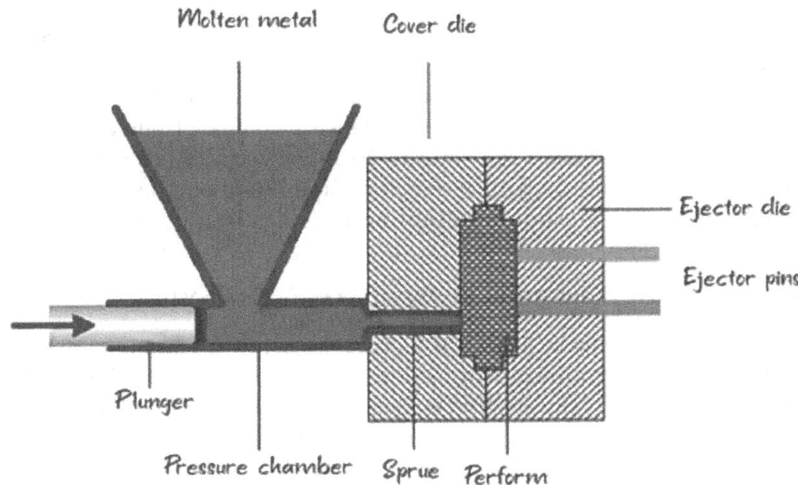


**FIGURE 3.6** Pressure die infiltration.

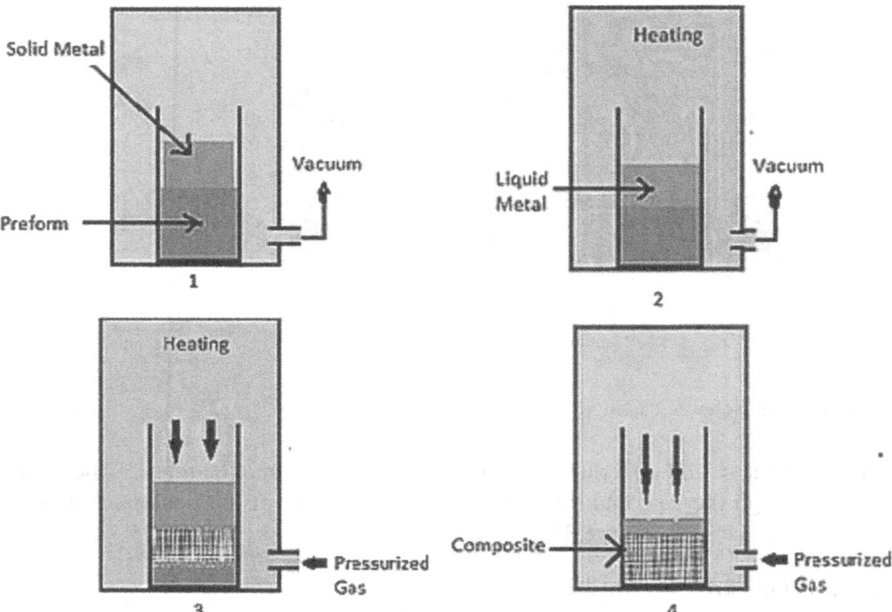


**FIGURE 3.7** Gas pressure infiltration.

Pneumatic penetration method is used to manufacture large composite parts. Due to the short contact time of the fibre with the molten iron, this method allows the use of uncoated fibres. Compared with the method using mechanical force, air pressure penetration leads to a low fibre damage rate.

#### 3.4.1.5 Infiltration

Infiltration is a liquid composite process in which the pre-formed dispersed layer (ceramic pellets, fibres, braids) is submerged in a molten matrix metal which really fills the gap between the dispersed phase materials. The strength of the infiltration mechanism can be the capillary energy of the dispersed phase (spontaneous infiltration) or the external pressure (gaseous, hydraulic, electrical, centrifugal or ultrasonic) added to the liquid matrix phase (mandatory infiltration). Infiltration is one of the techniques used to prepare composite materials of copper tungsten.

### 3.4.2 Solid State Fabrication of Metal Matrix Composite

Solid-state manufacturing of MMCs is a process where the formation of MMCs is the result of matrix metal and dispersed phase inter-diffusing at higher temperature and pressure in a solid state. The temperature range of the solid manufacturing method (compared to the liquid processing of MMCs) eliminates adverse effects at the boundary between the matrix and the distributed (reinforced) form. Following sintering, the MMC substance may often be deformed by grinding, welding, rubbing, stretching or crushing. The deformation could be either cold (underneath

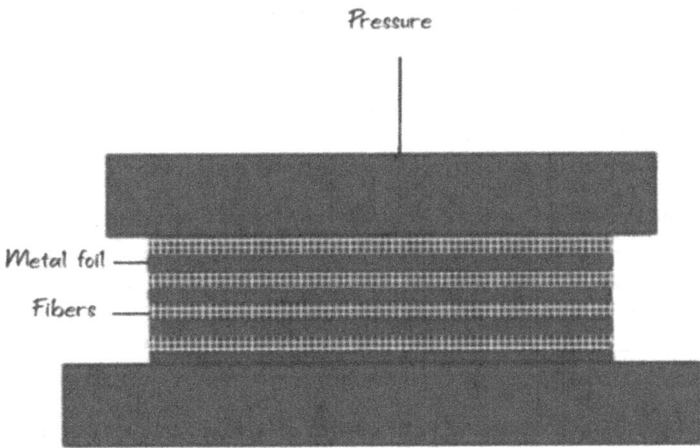


**FIGURE 3.8** Diffusion bonding.

the recrystallization temperature) or hot (above its recrystallization temperature). Deformation of sintered matrix composites with scattered process based on short fibres leads to the desired alignment of the fibres and anisotropy of the properties of the material (improved strength in the direction of the fibres).

There are two main groups of solid-state fabrication of MMCs:

1. Diffusion bonding.
2. Sintering.

#### 3.4.2.1 Diffusion Bonding

Diffusion bonding is indeed a solid-state process in which a matrix in the form of a foil and a diffuse layer in the form of long fibres are placed in a proper pattern instead of being pressurized at high temperature. The resulting laminated plastic material appears to have a multi-layer construction, as shown in Figure 3.8. Diffusion bonding is used for the manufacture of simple shaped parts (plates, tubes). Types of diffusion bonding include roll bonding and wire/tissue winding. Roll bonding is a method of rolling (hot rolling or cold rolling) sheets of two separate metals (such as steel and aluminium alloys) together, leading to the creation of a laminated matrix with a metallurgical link between the two. Wire/fibre winding is a method in which solid ceramic fibres with metal wires are mixed instead and forced at high temperatures.

#### 3.4.2.2 Sintering

Sintering of MMCs is a procedure wherein matrix metal powder as well as short fibres is blended with a dispersed phase powder for consequent compaction as well as sintering (sometimes liquid) of the solid state. Sintering is a process that consolidates metal powders by dehydrating the "green body" dense content to an elevated temperature below melting temperature, whenever the separated proton material diffuses into adjoining metal powder. The sintering method can obtain materials containing up to 50% of the dispersed phase compared with the liquid manufacture of MMCs.

The following manufacturing methods use sintering and deformation operations in combination:

1. Hot pressing fabrication of MMCs.
2. Hot isostatic pressing fabrication of MMCs.
3. Hot powder extrusion fabrication of MMCs.

#### 3.4.2.3 Hot Pressing Fabrication of Metal Matrix Composites

In this manufacturing process, sintering is performed under unidirectional pressure applied by hot press. The manufacturing stages of the hot pressing process can be divided into three: first, in the mould-filling stage, the powder is placed in the mould; then, in the second stage hot press the pressed parts in the mould; and in the final stage, the part is ejected because the upper punch moves upward to eject the part from the die.

#### 3.4.2.4 Hot Isostatic Pressing Fabrication of Metal Matrix Composites

In this manufacturing process, the liquid or gaseous medium surrounding the compressed part is sintered at high temperature under pressure exerted from multiple directions.

#### 3.4.2.5 Hot Extrusion Fabrication of Metal Matrix Composites

In this process, sintering is carried out under the pressure applied by the extruder at high temperature. It is of two types, namely forward extrusion and backward extrusion. A typical diagram of direct and indirect extrusion is illustrated in Figure 3.9, and Figure 3.10 illustrates the hot temperature extrusion, both direct and indirect.

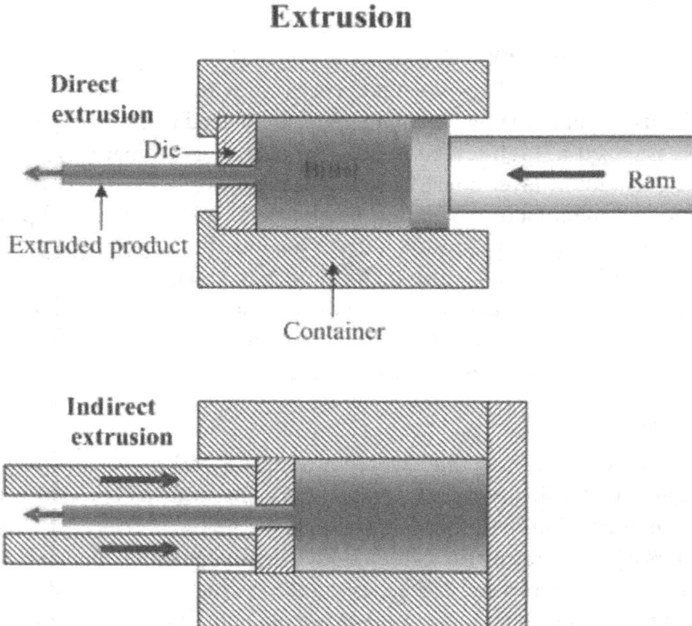


**FIGURE 3.9** Extrusion. (a) Direct and (b) indirect.

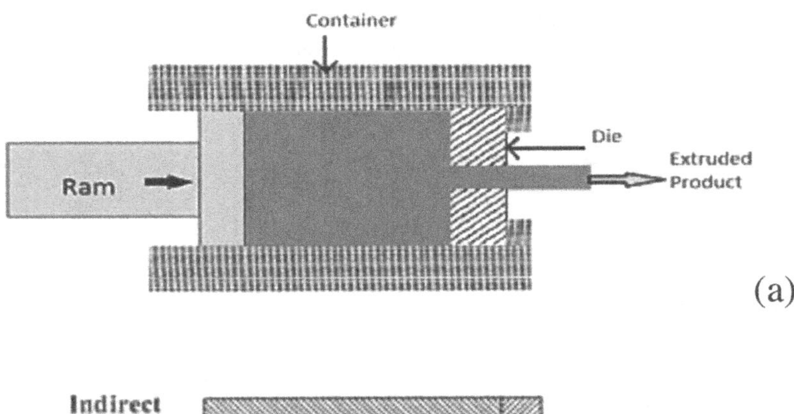


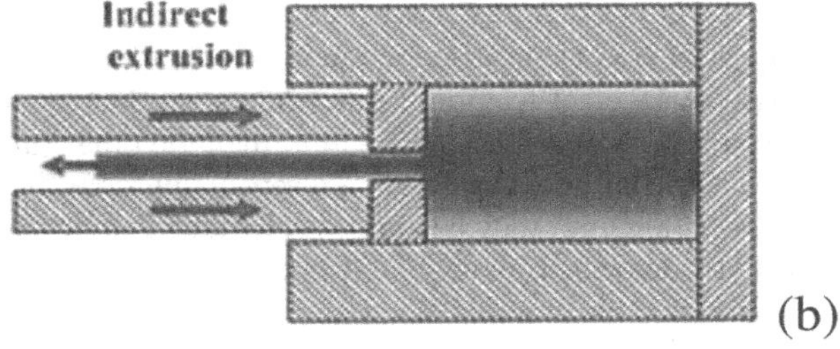


**FIGURE 3.10** Hot temperature extrusion. (a) Direct and (b) indirect.

## 3.5 WEAR THEORY

Wear in object occurs whenever two surfaces in relative movement engage with one another. Wear could be characterized by a gradual removal of material towards the contact region through relative motion. Experts have established several wear hypotheses that take into consideration the thermal and mechanical properties and environmental factors of the kernel (such as the strength of the frictional body and the tension of the touch area). Throughout 1940, Holm measured the volume of worn content on the unit's slipping path from the atomic process of wear.

Kragelski put forward the theory of wear fatigue. Due to irregularities in the solids, their interaction on the slide is discrete, and the various locations where they touch together form the actual contact area. Under the action of normal force, these irregularities penetrate or flatten each other, and the corresponding stress and strain rise in the actual contact area. During the sliding process, the specified amount of content is sensitive to multiple repetitions, which weakens the material and eventually causes cracking.

### 3.5.1 Types of Wear

Wearing issues have been a major concern for the most fundamental wear tests, and the so-called dry friction has been investigated in order to mitigate the effects of fluid lubricants. Dry friction is described as friction that is not intentionally lubricated, but it would be recognized as friction under the lubrication of atmospheric gases, in particular oxygen. There are five different kinds of wear:

1. Abrasive.
2. Adhesive.
3. Erosive.
4. Surface fatigue.
5. Corrosive.

#### 3.5.1.1 Abrasive Wear

The wear arising from the hard surface slipping to the soft surface and cutting the groove from the softest material may be described as abrasive wear. In fact, this might be the case of most of the failures. Abrasive wear can be caused by grinding, rough particles or the abrasion of one of the friction surfaces. This hard material may start from one of the two surfaces of friction. In the sliding mechanism, wear may be induced by the unevenness on one surface (if harder than the other surface) and the generation of wear debris. These wear debris are repeatedly deformed, so they are work hardened for oxidation until they become, harder than two or anyone. These wear debris are repeatedly deformed, so they are work hardened for oxidation until they become, harder than two or anyone. When any surface (usually harder than the second one) extracts the surface from the second surface, two-piece abrasive wear occurs (Figure 3.11). However, this mechanism usually changes to three-piece abrasive wear because the wear debris will serve like an abrasive here between two airfoils (Figure 3.11). The abrasive can act as an abrasive in the shell in which the abrasive is fixed proportional to the base, or as an abrasive that produces a series of dents (rather than scratches) when the abrasive rolls. According to recent tribological

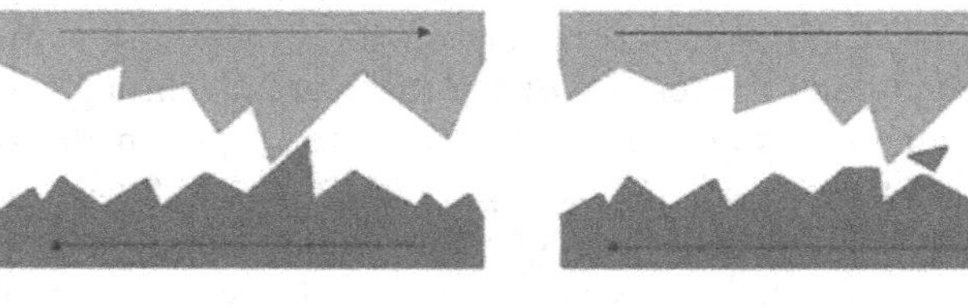


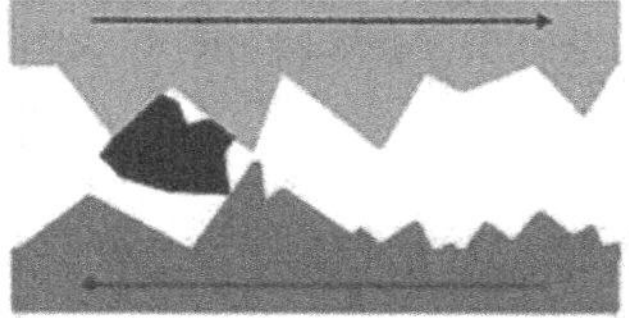

**FIGURE 3.11** Abrasive wear.

surveys, in industrial practice, abrasive wear is responsible for the largest amount of material loss.

#### 3.5.1.2 Adhesive Wear

Adhesive wear could be described as wear caused by the material exchange between different surfaces or by the setback of any such surface due to local bonding between the solid surfaces in contact (Figure 3.12). In order to wear the adhesive, the surfaces must be in intimate interaction with one another. Separate surfaces, such as lubricating film and oxide film, adversely affect the ability for adhesion.

#### 3.5.1.3 Erosive Wear

Erosive wear is described as the cycle of metal removal owing to positive particles reaching the surface. Erosion is induced by gas or liquid which might or might not have transported solid fragments on the surface. If the angle of impact is high, the resultant wear is quite close to wear. When the impact angle is perpendicular to the surface, the material will shift due to plastic flow or fall off due to brittle failure. A typical erosive wear mechanism is illustrated in Figure 3.13.

#### 3.5.1.4 Surface Fatigue

Strong surface wear is induced by cracks triggered by surface fatigue. The term "fatigue" is commonly employed in the case of collapse, where cyclic loading of solids requires stretching and compressing over a certain essential tension. Repetitive loading produces microcracks beneath the surface at the site of the normal weak points. Macrocracks can occur throughout the ensuing loading/unloading. Once the crack approaches the critical point, it will switch its orientation to appear on the surface so that it will peel off the flat plate-like particles during the wear cycle. The extent of the stress cycles needed to induce these failures declines as the stress rises. Vibration is the primary source of fatigue and wear. A schematic diagram of surface fatigue wear mechanism is illustrated in Figure 3.14.

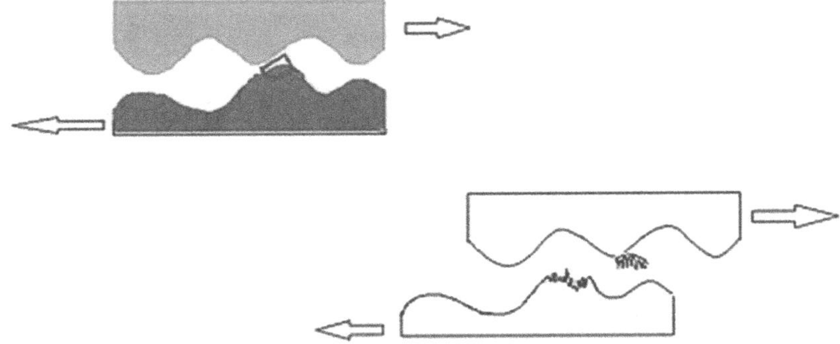

**FIGURE 3.12** Adhesive wear.

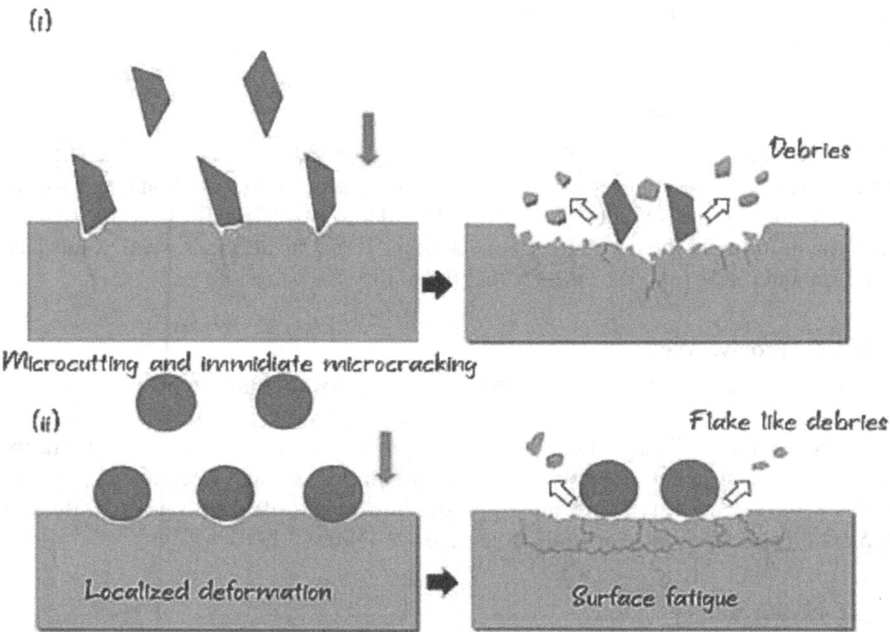


**FIGURE 3.13** Erosive wear.

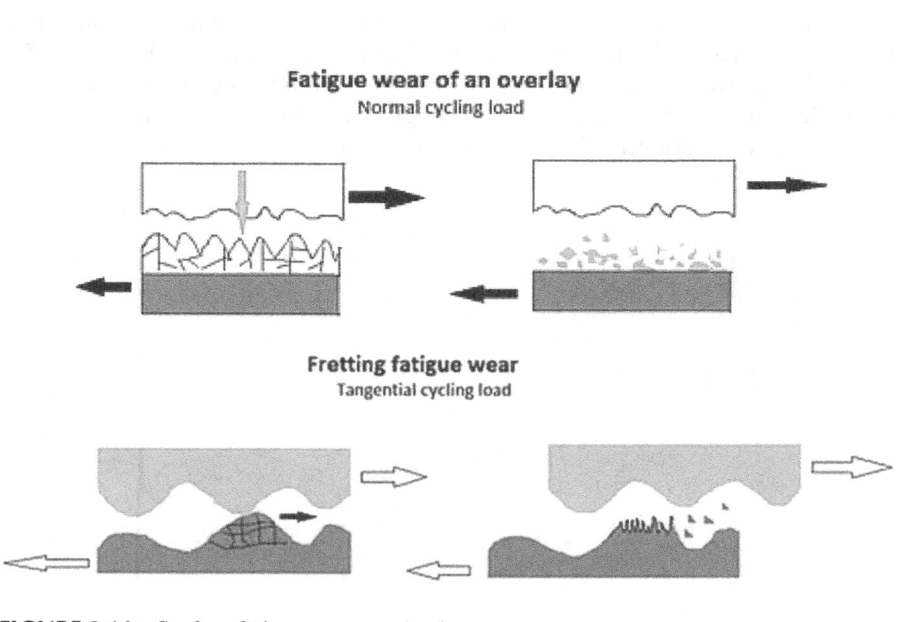


**FIGURE 3.14** Surface fatigue wear mechanism.

#### 3.5.1.5 Corrosive Wear

Often metals, such as aluminium, are thermodynamically reactive in air and interact with oxygen to create oxides. When the interface bond between them is weak, a layer or scale is usually formed on the surface of the metal or alloy. Corrosion and abrasion are due to the influence of the atmosphere, acid, gas, alkali, etc., and they gradually erode or deteriorate the unprotected metal surface. This kind of wear produces cracks and perforations and ultimately dissolves metal pieces.

## 3.6 LITERATURE SURVEY

Azimi et al. (2015) analyzed the optimizing consolidation behaviour of Al7068 with TiC conceptualized by the Taguchi experiment. The results show that 30 min hot pressing under a pressure of 500 MPa at 385 °C provides a high compressive strength and hardness of 938 MPa and HV 265 respectively. Additional experimental results showed that increased pressure has no obvious impact on mechanical properties and that further temperature increase had a reverse effect caused by heavy grain growth.

Zeeshan and Mangshetty (2018) studied the difference in Al property by the addition of different wt. % of SiC and $Al_2O_3$ reinforcement, such as SiC 2%+$Al_2O_3$ 8% and SiC 4%+$Al_2O_3$ 6% by stir casting method. The results show that reinforced composite is superior to base alloy7068 in terms of tensile strength and hardness. Maximum tensile strength and hardness was observed on SiC 6% +$Al_2O_3$ 4%.

Sajjan et al. (2017) observed Al7068 reinforced with varying graphite % (such as 5, 10 and 15%) prepared by stir casting method. The experiment result shows that the mechanical properties are increased with the % of graphite up to 10% and and then decreases gradually. It could be owing to an inadequate mixture of the reinforcement in the matrix. The improvement in weight proportion of graphite reinforcement resulted in a large rise in hardness and resistance to wear. As the percentage of graphite reinforcement enhanced, the rate of wear was reduced dramatically and therefore the wear resistance of the sample boosted.

Sharifi et al. (2011) analyzed the mechanical and wear conduct of aluminium metal matrix A coated with boron carbide. Increase in the amount of boron carbide nanoparticles enhances the hardness value and ultimate tensile strength and decreases the ductility, and this has been observed in a sample comprising $B_4C$ 15%. At a specific sliding distance, the friction coefficient of nanocomposites decreased and appeared to achieve a steady state.

Siddaram and Purohit (2011) evaluated optimization of the mechanical properties of aluminium metal matrix composite of grade 7068 containing a varying percentage of reinforcement such as Al7068+SiC 5%+Tur Husk 5%, Al7068+SiC 5%+Tur Husk 10% and AL7068+SiC 5%+Tur Husk 15% processed by the stir casting process. Tensile strength and hardness value increase after increasing the weight percentage of tur husk. However, a further increase in tur husk decreases the tensile strength of AA7068. Meanwhile, tensile strength increases and hardness value decreases after increasing the stirring time and pouring temperature. Also, wear property increases after increasing the pouring temperature but further increase of this temperature decreases the wear of AA7068.

Abdollahi et al. (2014) analyzed the dry sliding tribological characteristics and mechanical characteristics of the Al2024 and $B_4C$ nanocomposites generated by mechanical milling and hot extrusion. Hardness and strength of Al2024 alloy increase due to the addition of 5% $B_4C$ by the process of mechanical milling. The rise in the barrier of dislocation motion limits their transportation.

Subramaniam et al. (2018) evaluated the mechanical properties of Al7075 with reinforcement such as $B_4C$ and coconut shell fly ash manufactured by stir casting. Al7075 alloys were fabricated with a different weight percentage of $B_4C$ such as 3, 6, 9 and 12% and fixed 3% weight of coconut shell fly ash. The hardness of composite increased by 33% by the addition of reinforcement $B_4C$ 12%. Tensile strength of composite increased by 66% by the addition of 9 wt. % boron carbide. Further addition of reinforcement decreases the tensile strength.

Monikandan et al. (2018) have shown that hybrid composites have self-lubricating properties and can become resource-saving materials. AA6061 alloys, 10 wt. %, were reinforced with 2.5, 5 and 7.5 wt. % of $B_4C$ and $MoS_2$ hybrid composite. $MoS_2$ particles with a various wt. % were produced using stirred casting technology and evaluated for their mechanical and tribological properties. A dry sliding tribology study using a pin-bottom triple booster under atmospheric conditions shows that an $MoS_2$-lubricated friction layer is formed on the worn needle surface. This has a substantial influence on friction performance. The addition of $MoS_2$ particles reduces hybrids. Delamination and wear are observed as a mechanism to control wear and are in the form of flaky fragments, forming flow-type debris, and are also observed on the wear sales surface of hybrid composites.

Baradeswaran and Perumal (2014) investigated the mechanical as well as wear properties of Al7075 with reinforcement such as aluminium oxide and graphite. Fabrication was done by the stir casting process. Hybrid composites lead to increased $Al_2O_3$ ceramic particles. Increased hardness of hybrid composites, with the increase of $Al_2O_3$, is higher than the base alloy. Adding $Al_2O_3$ particles will increase tensile strength, compressive strength and flexural strength. The composite material is harder than even the base alloy. In addition, graphite has been commonly known for reducing aluminum alloy stiffness, tensile strength, compressive strength and flexural strength and for adding $al_2O_3$ to the hybrid structure. The existence of graphite in hybrid composites tends to maintain and reduce wear due to the generation of a thin coating of graphite upon this friction surface. Domination Costume The mechanism is wear and tear. Graphite coefficient may also be reduced in hybrid composites. Composite friction is attributed to the reason that graphite acts as a solid lubricant mostly during wear test.

Kumar et al. (2010) studied the Al6061 with SiC and Al7075 with $Al_2O_3$ on the composite metal matrix's mechanical as well as tribological features. Composites are produced through liquid metallurgy procedures, with reinforcement weight of 2–6% in the matrix material. Pradhan and Dehari (2019) studied the influence of electrical discharge machining (EDM) parameters on machinability for the Al7075–$B_4C$ and TiC hybrid and used ELECTRE optimization technique to obtain the best combination. Shukla et al. (2018) studied die-sink EDM machining of Al-LM6/SiC/$B_4C$ hybrid composites to evaluate its influence of process parameters on different responses.

In this research, the density of composite is found to be improved than the base matrix. The hardness of composite is found to be improved by raising the volume fraction in Al6061–SiC and Al7075–$Al_2O_3$, ranging from 60 to 97 VHN and 80 to 109 VHN, respectively. The strength and wear resistance of the Al6061 with SiC are higher than the Al7075 with the $Al_2O_3$ composite. It can be inferred from the analysis that the Al6061–SiC had outstanding mechanical and tribological properties.

Nazik et al. (2016) analyzed AA7075 composite reinforced with different weight % of $B_4C$ bearing 2, 4, 6 and 8% by weight of the materials produced by the powder technology. The hardness test showed that the hardness increased after 6 wt. % of boron carbide was added to the reinforcement. Wear check evaluation has shown that the integration of 10% $B_4C$ to the overall weight of the composite has greatly improved wear resistance and that the wear rate is even lower with the material mixture of 11% $B_4C$ and 5% graphite. There exists wear analysis performed in Al composites comprising specific reinforcements: a sample containing up to 8% $B_4C$, 8% SiC and 4% $Al_2O_3$ has the maximum friction coefficient for even a load of 20 N and another sample containing up to 7% SiC and 6% $Al_2O_3$ does have the lowest friction coefficient for a time span of 360 seconds.

Muthazhagan et al. (2014) studied the influence of Al6061 with the reinforced boron carbide and graphite. This research survey deals with the mechanical properties change with the addition of reinforcement, and it is fabricated through stir casting method. Hardness test of the Al–$B_4C$ and graphite was carried out using Brinell hardness machine and tensile strength test was carried out by UTM machine. It is uniformly distributed in both boron carbide and graphite. It was identified that even a proportion of the graphite content enhances the ductility and decreases the tensile strength. On the other hand, the addition of boron carbide increases the hardness of the composite. The heat-treated composites were polished and observed in a scanning electron microscope. The micrograph of Al6061 with reinforcing agent shows an even distribution of graphite particles and boron carbide. It is also observed that porous site was reduced.

Murugan et al. (2018) studied the mechanical properties of $Al_2O_3$ and SiC reinforced Al6061-T6 hybrid matrix composite. It was fabricated through stir casting method, keeping alumina 7 wt.% constant and increasing silicon carbide percentage as 10, 15 and 20%. Upon manufacturing, the sample was prepared and characterized to determine the different mechanical properties, such as the tensile, compressive and impact strength of the material. From the observation, it was noticed that tensile strength, maximum yield strength and maximum fracture stress value got maximum at 20 wt.% of SiC. Impact strength was found to be increased after addition of Sic. With increase in the quantity of SiC in the matrix, ceramics can establish a close bond with the Al6061 content and thereby enhance the properties of composites.

Pradhan (2020) examined the mechanical, tribological behviour and machinability using EDM on a mixture reinforced with varying formulations of SiC and glass particulates produced by stir casting method. The role of composition on responses was also identified with machine parameters. The optimum EDM combination was implemented by solving the MULTI-MOORA method integrated with PCA method. Glass particles do not really relate to tensile strength; yet, they improve the tribological properties substantially. The composition, as well as pulse current, was the most important parameter. The constitution resulted in the best machinability configuration.

## 3.7 OPTIMIZATION PROCESS

In this chapter, wear test was conducted on aluminium hybrid composite, i.e. (Al7068 + 2% $B_4C$ + 1% alumina) and (Al7068 + 2% $B_4C$ + 2% alumina), and Al7068 alloy using pin-on-disk machine. Wear rate and coefficient of friction (COF) were calculated to find the effect of $B_4C$ and $Al_2O_3$ on the tribological behaviour of the materials. Further, the DEMATEL analysis was conducted to assess the connection between cause and effect, as well as the aspect that most affects the process.

### 3.7.1 DEMATEL Method

Decision making trial and evaluation laboratory (DEMATEL) was first developed by the Geneva Research Centre of Battelle Memorial Institute for visualizing structure of complex causality matrix or graph. Modelled as a structured method, it is especially useful when analyzing causes and solutions affect. The relationship between system components of DEMATEL may show the mutuality between components and assist in the evolution of maps to improve the relative relationship among them; it can also be used for surveys and solve complex problems. This method not only transforms the interdependence into causal grouping by matrix but also find the key influencing factors of complex structural systems.

Because of its usage and potential, DEMATEL gained a lot of attention for a few years. A systematic analysis of the new DEMATEL-based literature has also been performed in this study. Several scholars have used it to address dynamic machine issues in diverse fields. DEMATEL has been applied to make smarter choices in other settings since certain real-world applications contain unreliable and ambiguous details, but for the sake of education, no thorough analysis of this technology has been implemented.

Multi-criteria analysis using the DEMATEL method consists of the following steps:

Step 1: Form a set of direct-influence matrix [Z]. To grasp the link between factors $F = F_1, F_2, \ldots, F_n$ within the system, suppose $l$ experts within the decision group $E = E_1, E_2, \ldots, E_n$ are required the employment of integer value to indicate the direct influence that factor $F_i$ has on factor $F_j$, using an integer scale of 'no influence (0)', 'low impact (1)', 'medium impact (2)', 'high impact (3)' and 'very high impact (4)'. Then, the individual direct-impact matrix $Z_k = [z_{ij}^k]_{n*n}$ can be formed by the $k^{th}$ expert, in which all central diagonal components are equal to zero and $z_{ij}^k$ represents decision the judge of the degree of influence maker of the factor $F_i$ factor. Through outlining the expert views, the community has a clear impact matrix $Z_k = [z_{ij}]_{n*n}$, which can be obtained by

$$Z_{ij} = \frac{1}{l}\sum_{k=1}^{l} Z^k \quad i, j = 1, 2, .., n \tag{1.1}$$

Using equation 1.1, we get the following direct influence matrix [Z]

$$[Z] = \begin{bmatrix} 0 & 3 \\ 4 & 0 \end{bmatrix}$$

Step 2: Set up a comprehensive significant-impact matrix. Once the group directly impacts the matrix [Z], the normalized direct-impact matrix is used as $X = [x_{ij}]_{n*n}$, which can be achieved by using

$$X = \frac{z}{s} \tag{1.2}$$

$$S = \max\left(\max_{1\le i\le n}\sum_{j=1}^{n} Z_{ij},\ \max_{1\le i\le n}\sum_{j=1}^{n} Z_{ij}\right) \tag{1.3}$$

Using equations 1.2 and 1.3, we get the following direct influence matrix [X]

$$[X] = \begin{bmatrix} 0 & 0.75 \\ 1 & 0 \end{bmatrix}$$

Step 3: Create a complete effect matrix [T]. Using its normalized explicit influence matrix [X], the complete-impact matrix $T = [t_{ij}]n*n$ is then calculated by adding indirect impact.

$$T = X + X^2 + X^3 + \ldots.. + X^h = X(I - X)^{-1}\text{, where } h \to \infty \tag{1.4}$$

where $I$ represents an identity matrix.

Therefore, using equation 1.4, we get the following total impact matrix [T]

$$[T] = \begin{bmatrix} 3 & 3 \\ 4 & 3 \end{bmatrix}$$

Step 4: Compute the vectors $R$, which portrays the sum of the rows, and $C$, which indicates the sum of the columns, of the total impact matrix represented by the following equations:

$$R = [r_j]_{n\times 1} = \left[\sum_{j=1}^{n} t_{ij}\right]_{n\times 1} \tag{1.5}$$

$$c = [c_j]_{1\times n} = \left[\sum_{i=1}^{n} t_{ij}\right]_{1\times n}^{T} \tag{1.6}$$

Table 3.1 shows the $R_i$ and $C_i$ values using equations 1.5 and 1.6.

**TABLE 3.1**
**$R_i$ and $C_i$ for matrix [$T$]**

| | Wear rate | COF | $R_i$ |
|---|---|---|---|
| Wear rate | 3 | 3 | 6 |
| COF | 4 | 3 | 7 |
| $C_i$ | 7 | 6 | |

**TABLE 3.2**
**$R_i - C_i$ and $R_i + C_i$**

| Criterion | $R_i$ | $C_i$ | $R_i + C_i$ | $R_i - C_i$ | Cause and effect |
|---|---|---|---|---|---|
| Wear Rate | 6 | 7 | 13 | −1 | Effect |
| COF | 7 | 6 | 13 | 1 | Cause |

Step 5: Find the $R_i - C_i$ and $R_i + C_i$ values for each criterion. $R_i + C_i$, named 'protrusion', demonstrates the strength of impacts such factor receives. In other words, $R_i + C_i$ represents the degree to which this factor plays a central role within the system. Similarly, $R_i - C_i$, called 'relationship', indicates that the overall effect of its factor contributing to the mechanism will be classified into the category in consequence. If $(R_i - C_i)$ is positive, the $F_j$ criterion has a net impact but in the other factors and could be categorized into a class of causes; if $(R_i - C_i)$ is negative, then the $F_j$ criteria is affected by the other variables as a whole rather than could be categorized under the group of results. Table 3.2 shows the $R_i - C_i$ and $R_i + C_i$ values by adding the $R_i$ and $C_i$ values for each criterion.

On the basis of the calculations carried out, it can be concluded that both wear rate and COF are dominant factors, as the $R_i + C_i$ value is same for both the criteria. Also, the $R_i - C_i$ value for the COF is positive, which puts it into the cause group; and the wear rate is in the effect group since its $R_i - C_i$ value is negative. This shows that during the wear test of the material, the COF affects the wear rate of the material.

## 3.8 RESULT AND DISCUSSION

### 3.8.1 Effect of Wear Parameters on Wear Rate and COF

The rates of wear and COF are registered from pin-on-disk machine, as depicted in Table 3.3.

#### 3.8.1.1 On Wear Rate

The Taguchi methodology is a quantitative approach to the analysis of means and signal-to-noise (S/N) ratios. It includes graphic designing the basic impact of various causes and the visual development of the essential influences. The S/N ratio is shown in Figure 3.15, which explores the effects of the wear behaviour on the

**TABLE 3.3**
**Experimental result for the wear test**

| Exp. no. | Alumina wt.(%) | Normal Load (N) | Sliding Speed (RPM) | Wear Rate ($mm^3/s$) | COF |
|---|---|---|---|---|---|
| 1 | 0 | 10 | 400 | 785 | 0.125 |
| 2 | 0 | 20 | 500 | 795 | 0.119 |
| 3 | 0 | 30 | 600 | 800 | 0.115 |
| 4 | 1 | 10 | 400 | 650 | 0.102 |
| 5 | 1 | 20 | 500 | 665 | 0.100 |
| 6 | 1 | 30 | 600 | 680 | 0.096 |
| 7 | 2 | 10 | 400 | 520 | 0.088 |
| 8 | 2 | 20 | 500 | 540 | 0.081 |
| 9 | 2 | 30 | 600 | 560 | 0.075 |

response material removal rate. The effect of alumina (wt.%), normal load and sliding speed on the rate of wear can be seen in the figure. The below graph shows that as the value of alumina (wt.%) increases, wear rate decreases. Thus, the minimum wear rate occurs at highest value (wt.%) of aluminium oxide, i.e. at 2%. But the rate of wear rises when the value of the sliding speed increases. The graph shows that the rate of wear is at the highest value of 600 RPM. Also, with an increase in the value

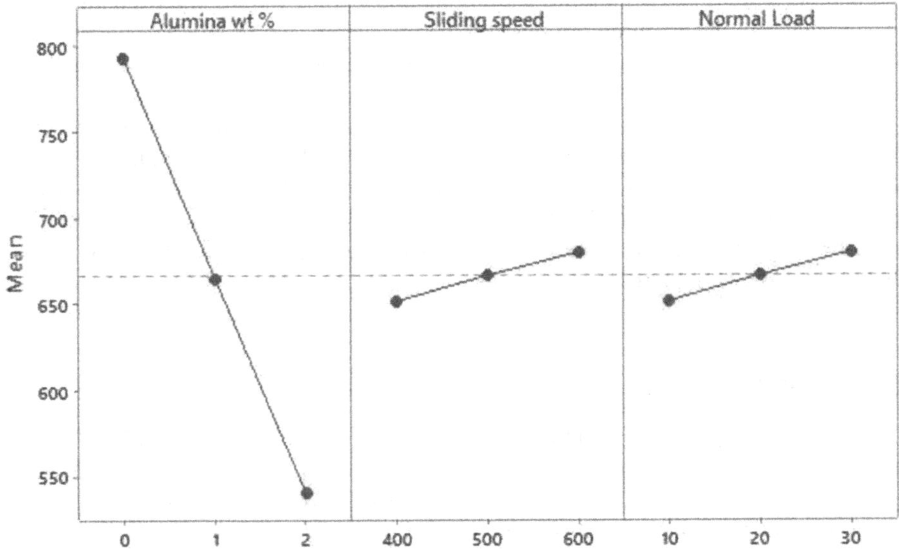


**FIGURE 3.15** Main effect plot for wear rate.

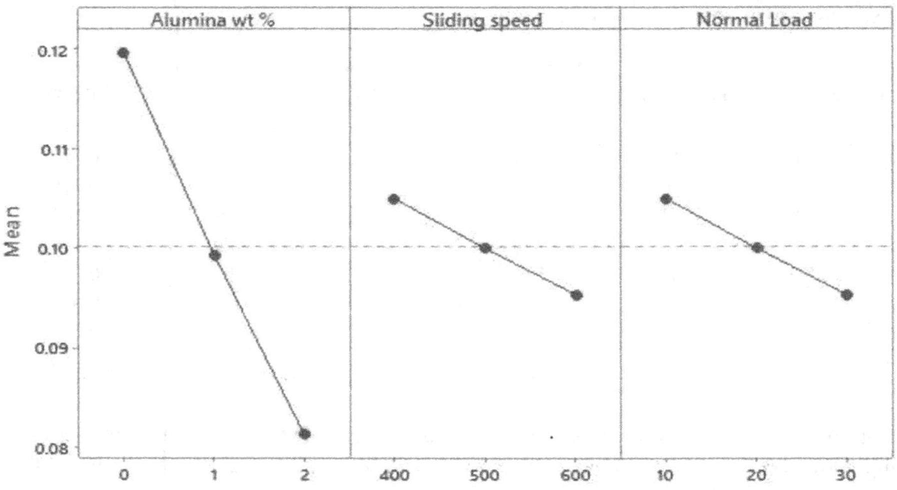


**FIGURE 3.16** Main effect plot for COF.

of the normal load (N), the rate of wear increases and reaches a maximum of 30 N. Figure 3.15 shows the main effect plot for the rate of wear.

#### 3.8.1.2 On Coefficient of Friction

The Taguchi methodology is a quantitative approach to the study of means and S/N ratios. It includes graphic designing the basic impact of various causes and the visual development of the essential influences.

The S/N is shown in Figure 3.15 to test the impact of wear parameters on material removal rate using Minitab software

The effect of alumina (wt.%), normal load and sliding speed on COF could be seen in Figure 3.16. The figure indicates that as the value of alumina (wt.%) increases, COF decreases. Therefore, the minimum COF is at highest value (wt.%) of alumina ($Al_2O_3$), i.e. at 2%. The COF also decreases when the value of sliding speed increases. The graph shows that the COF reaches a minimum value at 600 RPM. Also, with an increase in the value of normal load (N), the COF decreases and reaches the minimum at 30 N.

## 3.9 CONCLUSION

The present study investigated the wearability of a hybrid compound (Al7068 + $B_4C$ + alumina [$Al_2O_3$]) on pin-on-disk machine and also evaluated the influence of wear parameters on wear rate and COF in a hybrid composite. The experiments were run under such a wide scope of conditions: aluminium oxide wt.%, sliding velocity and standard load. In the Taguchi process, Minitab 18 software is being used to characterize and evaluate the impact of the wear behaviour upon its response. It is clear that as the value of alumina ($Al_2O_3$) (wt.%) increases, wear rate decreases. Thus, the minimum wear rate is at highest value (wt.%) of aluminium oxide, i.e. at 2%. But the wear rate increases when the value of sliding speed increases. The graph (Figure 3.15) shows that the wear rate reaches a maximum value at 600 RPM. Also, with the increase in the value of normal load (N), the wear rate increases and reaches the maximum at 30 N. Figure 3.14 shows the main effect plot for wear rate. Also, as the value of alumina ($Al_2O_3$) (wt.%) increases, COF decreases. Hence, the minimum COF is at highest value (wt.%) of alumina ($Al_2O_3$), i.e. at 2%. The COF also decreases when the value of sliding speed increases. The graph (Figure 3.16) shows that the COF reaches a minimum value at 600 RPM. Also, with an increase in the value of normal load (N), the COF decreases and reaches the minimum at 30 N. On the basis of the calculations carried out, it can be concluded that both wear rate and COF are the dominant factors, as the $R_i + C_i$ value is same for both the criteria. Also, the $R_i - C_i$ value for the COF is positive, which puts it into the cause group; and the wear rate is in the effect group since its $R_i - C_i$ value is negative. This shows that during the wear test of the material, the COF affects the wear rate of the material.

## REFERENCES

Abdollahi, A., Alizadeh, A., and Baharvandi, H. R. (2014). Dry sliding tribological behaviour and mechanical properties of Al2024–5 wt.% B4C nanocomposite produced by mechanical milling and hot extrusion. *Materials & Design*, 55:471–481.

Azimi, A., Shokuhfar, A., Nejadseyfi, O., Fallahdoost, H., and Salehi, S. (2015). Optimizing consolidation behaviour of Al 7068–TiC nanocomposites using Taguchi statistical analysis. *Transactions of Nonferrous Metals Society of China*, 25 (8):2499–2508.

Baradeswaran, A., and Perumal, A. E. (2014). Study on mechanical and wear properties of Al 7075/Al2O3/graphite hybrid composites. *Composites Part B: Engineering*, 56:464–471.

Kumar, G. V., Rao, C., Selvaraj, N., and Bhagyashekar, M. (2010). Studies on Al6061-SiC and Al7075-Al2O3 metal matrix composites. *Journal of Minerals & Materials Characterization & Engineering*, 9 (1):43–55.

Kumar, P. R., and Pradhan, M. K. (2016). Improvement in mechanical property of aluminium alloy LM-6 reinforcement with silicon carbide and boron carbide particles. *i-manager's Journal on Material Science*, 4:27–32.

Monikandan, V., Joseph, M., and Rajendrakumar, P. (2018). Application of full factorial design to study the tribological properties of AA6061-B4C and AA6061-B4CMoS2 composites. In *Proceedings of Asia International Conference on Tribology 2018*. Sarawak, Malaysia, September 2018.

Murugan, S. S., Jegan, V., and Velmurugan, M. (2018). Mechanical properties of SiC, Al2O3 reinforced aluminium 6061-T6 hybrid matrix composite. *Journal of the Institution of Engineers (India): Series D*, 99 (1):71–77.

Muthazhagan, C., Gnanavelbabu, A., Bhaskar, G., and Rajkumar, K. (2014). Influence of graphite reinforcement on mechanical properties of aluminium-boron carbide composites. *Advanced Materials Research*, 845:398–402.

Nazik, C., Tarakcioglu, N., Ozkaya, S., Erdemir, F., and Canakci, A. (2016). Determination of the effect of B4C content on density and tensile strength of AA7075/B4C composite produced via powder technology. *International Journal of Materials, Mechanics and Manufacturing*, 4 (4):251–261.

Pradhan, M. K. (2020). Tribological behaviour, machinability, and optimization of EDM of AA-2014 hybrid composite reinforced with SiC and glass particulates. In *Handbook of Research on Developments and Trends in Industrial and Materials Engineering*, volume 1, pages 228–269. IGI Global.

Pradhan, M. K., and Dehari, A. (2019). Optimization of process parameters for electrical discharge machining of AL7075, B4C and TiC hybrid composite using ELECTRE method. In *Optimization using Evolutionary Algorithms and Metaheuristics: Applications in Engineering*, volume 1, pages 57–80. CRC Press Taylor and Francis Groups.

Sajjan, B., Avinash, L., Varun, S., Varun Kumar, K., and Parthasarathy, A. (2017). Investigation of mechanical properties and dry sliding wear behaviour of graphite-reinforced Al7068 alloy. In *Technological Advancements in Materials and Manufacturing for Industrial Environment, volume 867 of Applied Mechanics and Materials*, pages 10–18. Trans Tech Publications Ltd.

Sharifi, E. M., Karimzadeh, F., and Enayati, M. (2011). Fabrication and evaluation of mechanical and tribological properties of boron carbide reinforced aluminium matrix nanocomposites. *Materials & Design*, 32 (6):3263–3271.

Shukla, M., Agarwal, P., Pradhan, M. K., and Dhakad, S. (2018). Experimental investigation of EDM parameters on Al-LM6/SiC/B4C hybrid composites. Applied Mechanics and Materials, 877:149–156.

Siddaram, B., and Purohit, G. (2011). A study on optimization of reinforcements SiC and tur husk on mechanical and tribological properties of AA 7068 MMC's by Taguchi..

Subramaniam, B., Natarajan, B., Kaliyaperumal, B., and Chelladurai, S. J. S. (2018). Investigation on mechanical properties of aluminium 7075-boron carbide coconut shell fly ash reinforced hybrid metal matrix composites. *China Foundry*, 15 (6):449–456.

Zeeshan, S. S., and Mangshetty, S. J. (2018). A study on mechanical and tribological properties of aluminum-7068 alloy reinforced with SiC and Al2O3. *International Journal of Technical Innovation in Modern Engineering & Science*, :478–488.

# 4 A Comparison of ACO and GA for Routing AGVs via C#

*Şahin Inanç*
Bursa Uludağ University
Bursa, Turkey

*Arzu Eren Şenaras*
Bursa Uludağ University
Bursa, Turkey

## CONTENTS



## 4.1 INTRODUCTION

Automated guided vehicles (AGVs) are used to transport all types of materials related to the manufacturing process at manufacturing areas (Fazlollahtabar and Saidi-Mehrabad, 2015). Material handling is an important part of the manufacturing systems. Due to the fact that this process is seen as a non-value-added process, material handling is tried to be automated today, and the performance of the enterprises is measured with this aspect. To automate material handling, AGV system comes first among many methods. This material handling system, which has been successfully implemented in many sectors, is becoming more and more relevant in our lives day by day, and by adding many other features on it, the usage area of AGV will increase.

## 4.2 A SHORT LITERATURE REWIEV

Haroun and Jamal (2015) compared the qualitative and computational performances of ant colony optimization (ACO) and genetic algorithm (GA). The study showed that ACO provides better results, especially with large problems.

Mukhairez and Maghari (2015) evaluated the performance of three algorithms (ACO, GA and simulated annealing [SA]) in solving the travelling salesman problem. In term of shortest distance between the cities, ACO fared better than GA and SA.

Şenaras et al. (2017) analyzed utilization rate of robot facilities with developed simulation model. Existent system is analyzed and then modelled in Arena 14.0 package program. Changes of capacity utilization rates are analyzed considering developed alternative suggestions.

Şenaras and İnanç (2018) analyzed the shortest way for establishing line of AGVs in an enterprise. A case study was developed via the dynamic programming method. Model was developed using Visual Basic for Application (VBA) in MS Excel. Thanks to this study, products are transported as soon as possible.

Islam et al. (2019) presented four different algorithms that can solve the traveling salesman problem and compared their performance. Dynamic programming gives the best result.

İnanç and Şenaras (2020) calculated the shortest way for the AGV line, which was found using ACO.

Valdez et al. (2020) made a comparison of GA and SA algorithms to determine which of these gives better results.

Şenaras and Şenaras (2020) developed an SD model for a flexible manufacturing system. System dynamics is a tool to test these rules without any production loss.

Mauluddin et al. (2020) made a comparison of ACO and GA. The authors illustrated that GA is the fastest algorithm for scheduling cases. This process gives a best solution effect speed to get the best scheduling solution.

## 4.3 ANT COLONY OPTIMIZATION

ACO was introduced by Dorigo and colleagues as a novel, nature-inspired metaheuristic for the solution of hard combinatorial optimization problems in the early 1990s (Dorigo and Blum, 2005). The ACO algorithm is an agent-based meta-heuristic that is used to find solutions to various optimization problems. Agents are called artificial ants (Dorigo et al., 1996).

## 4.4 GENETIC ALGORITHM

In the late 1960s, the concept of GA was first introduced by Holland and his colleagues (Holland, 1975). GAs are a family of computational models inspired by evolution. These algorithms encode a potential solution to a specific problem on a simple chromosome-like data structure, and apply recombination operators to these structures in such a way as to preserve critical information (Whitley, 1994:65).

## 4.5 COMPARING ACO AND GA FOR ROUTING AGVs

We compare the ACO and GA for calculating the shortest path of AGVs. AGV is usually used to transport goods from one spot to another in factory environments. These tools have predetermined roots. Twelve nodes were used in this implementation. The 12 nodes were used to detect the AGV's route. The distances of stations are shown as Table 4.1.

### 4.5.1 Ant Colony Optimization Algorithm

1. Start
2. Initialize parameters α, β, ρ, iterationCount, antCount
3. Set ants randomly on each node
4. Calculate tours for each ant
5. Update local pheromone for each tour
6. Update global pheromone for best tour
7. antCount = antCount – 1
8. iterationCount = iterationCount - 1
9. if antCount > 0 go to step 4
10. if iterationCount > 0 go to step 3
11. Display best result
12. Stop

Flow chart of ant colony algorithm application via C#. is shown as in Figure 4.1.

### 4.5.2 Genetic Algorithm

1. Start
2. Initialize Population
3. Calculate the fitness values of all chromosome
4. If maximum iteration reached go to step 8
5. Use crossing over method on chromosome
6. Use mutation method on chromosome
7. Go to step 3
8. Display best solution
9. Stop

Flow chart of genetic algorithm application via C# is shown as Figure 4.2.

Two programs have been developed for both the algorithms. The programs have been developed in C# programming language. The initial settings for the ant colony were set to alpha = 1 and beta = 1. 10,000 iterations were used for both algorithms. The systems used to run the programs are presented in Table 4.2. The results of the programs are given Table 4.3.

There are 12 stations and AGV will tour all stations.

**TABLE 4.1**
**Distances of Stations (Meters)**

| | Station1 | Station2 | Station3 | Station4 | Station5 | Station6 | Station7 | Station8 | Station9 | Station10 | Station11 | Station12 |
|---|---|---|---|---|---|---|---|---|---|---|---|---|
| **Station1** | 0 | 38.28 | 81.63 | 70.21 | 70.83 | 54.13 | 56.57 | 52.39 | 44.10 | 29.61 | 32.76 | 52.39 |
| **Station2** | 38.28 | 0 | 50.45 | 65.80 | 52.50 | 38.28 | 20.62 | 59.48 | 37.01 | 62.82 | 27.31 | 29.15 |
| **Station3** | 81.63 | 50.45 | 57.00 | 0 | 28.02 | 37.12 | 30.27 | 68.62 | 48.55 | 97.08 | 53.04 | 31.83 |
| **Station4** | 70.21 | 65.80 | 57.00 | 0 | 29.07 | 28.84 | 60.42 | 22.47 | 29.68 | 65.00 | 42.06 | 39.96 |
| **Station5** | 70.83 | 52.50 | 28.02 | 29.07 | 0 | 18.68 | 39.20 | 43.57 | 29.97 | 79.06 | 40.02 | 23.54 |
| **Station6** | 54.13 | 38.28 | 37.12 | 28.84 | 18.68 | 0 | 31.78 | 32.39 | 12.04 | 61.72 | 21.38 | 11.18 |
| **Station7** | 56.57 | 20.62 | 30.27 | 60.42 | 39.20 | 31.78 | 0 | 61.52 | 37.22 | 76.92 | 33.96 | 20.62 |
| **Station8** | 52.39 | 59.48 | 68.62 | 22.47 | 43.57 | 32.39 | 61.52 | 0 | 24.33 | 42.76 | 32.25 | 42.00 |
| **Station9** | 44.10 | 37.01 | 48.55 | 29.68 | 29.97 | 12.04 | 37.22 | 24.33 | 0 | 49.68 | 12.65 | 18.44 |
| **Station10** | 29.61 | 62.82 | 97.08 | 65.00 | 79.06 | 61.72 | 76.92 | 42.76 | 49.68 | 0 | 44.41 | 65.30 |
| **Station11** | 32.76 | 27.31 | 53.04 | 42.06 | 40.02 | 21.38 | 33.96 | 32.25 | 12.65 | 44.41 | 0 | 21.26 |
| **Station12** | 52.39 | 29.15 | 31.83 | 39.96 | 23.54 | 11.18 | 20.62 | 42.00 | 18.44 | 65.30 | 21.26 | 0 |

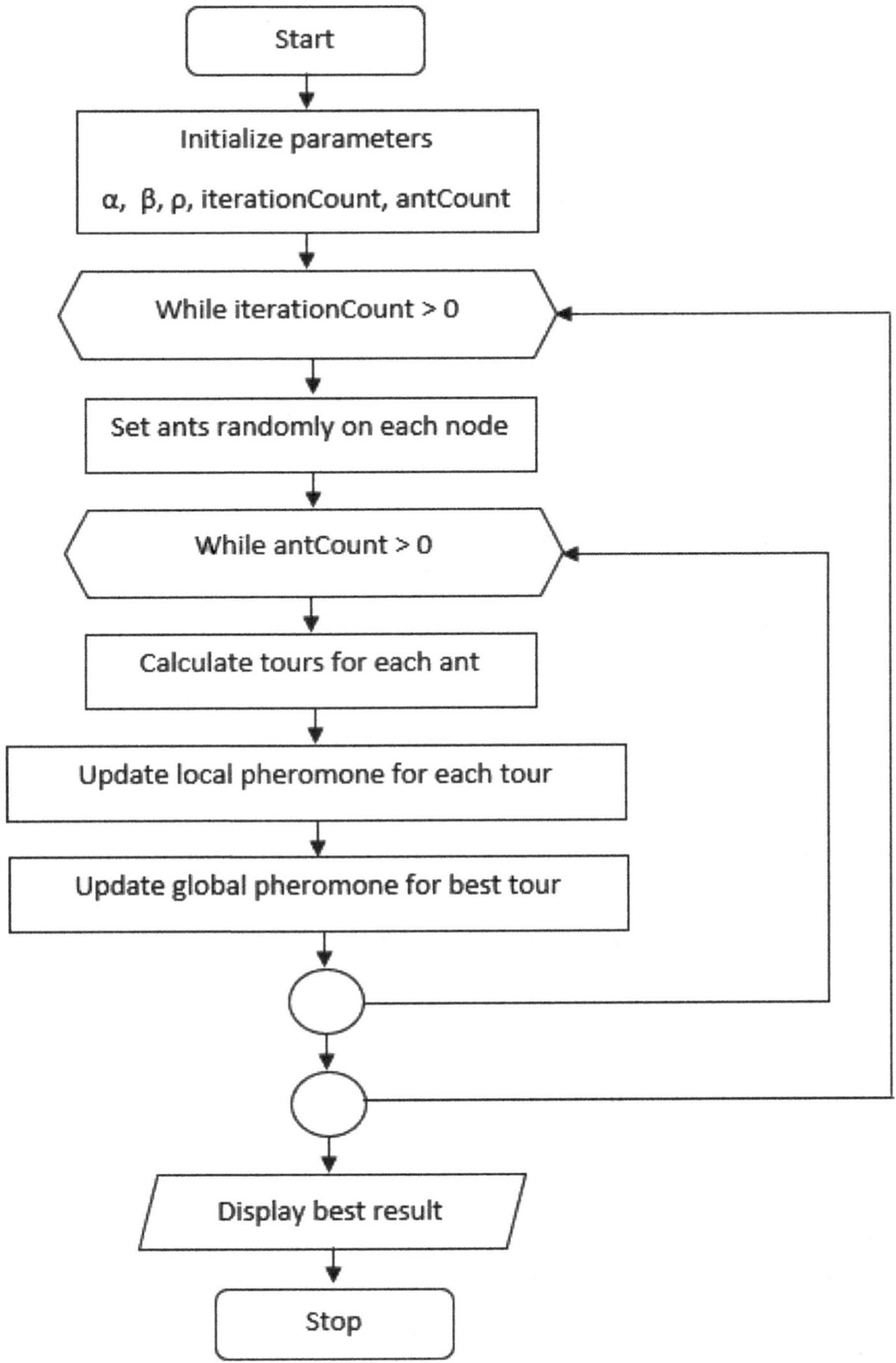


**FIGURE 4.1** Flow chart of ant colony algorithm application via C#.

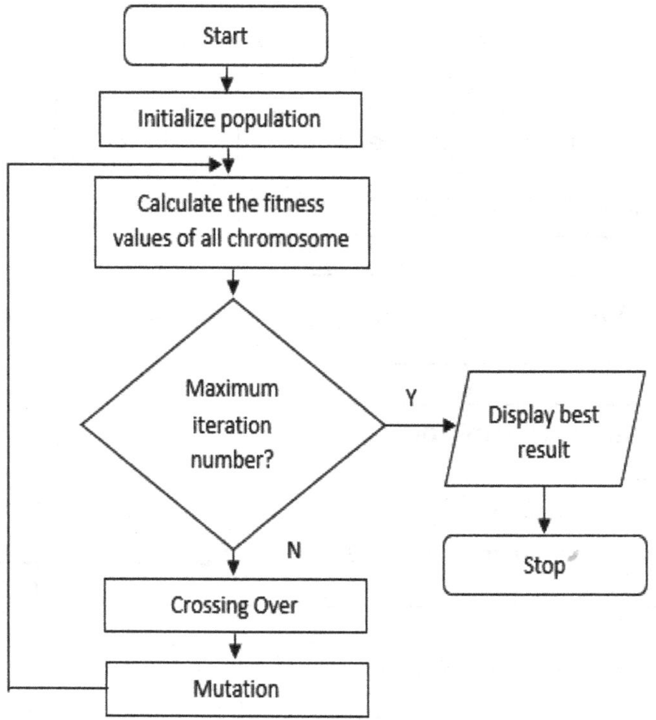


**FIGURE 4.2** Flow chart of genetic algorithm application via C#.

*The results for the ant colony algorithm in the calculations made with the programs are as follows:*

The shortest route: 1 - 10 - 5 - 4 - 8 - 9 - 12 - 6 - 11 - 3 - 7 - 2 - 1
Distance of route: 377, 80

*The results for the GA are as follows:*

The shortest route: 1 - 6 - 12 - 9 - 8 - 10 - 5 - 4 - 11 - 3 - 7 - 2 - 1
Distance of route: 443,24

**TABLE 4.2**
**Details of the system used for programs**

| **Processor** | **Inteli5** |
|---|---|
| RAM | 4 GB |
| Operating System | Windows 10 |

**TABLE 4.3**
**Results for each algorithm**

| | Time | Iteration |
|---|---|---|
| Ant Colony Optimization | 20.1 second | 10.000 |
| Genetic Algorithm | 4.8 second | 10.000 |

## 4.6 CONCLUSION

AGV line's shortest way was calculated via ACO and GA. Developed model was created in C# programming language. ACO is used for multiple AGV's path planning problems. Both the algorithms have their own advantages. While the ant colony algorithm finds better results, the GA can find faster results. As seen in other studies, GA can be more advantageous in different types of NP-hard problems. It can be said that ant colony algorithm can be more successful for solving traveling salesperson problems.

## REFERENCES

Dorigo, M., Maniezzo, V., & Colorni, A. (1996). "Ant system: optimization by a colony of cooperating agents," *IEEE Transactions on Systems, Man, and Cybernetics, Part B: Cybernetics,* 26:29–41.

Dorigo, M., & Blum, C. (2005). "Ant colony optimization theory: a survey," *Theoretical Computer Science,* 344(2-3):243–278.

Fazlollahtabar, H., & Saidi-Mehrabad, M. (2015). "Methodologies to optimize automated guided vehicle scheduling and routing problems: a review study", *Journal of Intelligent & Robotic Systems,* 77:525–545.

Haroun, S. A., & Jamal, B. (2015). "A performance comparison of GA and ACO applied to TSP," International Journal of Computer Applications, 117(20).

Holland, J. H. (1975). *Adaptation in natural and artificial systems.* The University of Michigan Press.

İnanç, Ş., & Şenaras, A. E. (2020). "AGV routing via ant colony optimization using C#," In: Kumar, K., Davim, J. P. (eds), *Optimization Using Evolutionary Algorithms and Metaheuristics.* Taylor & Francis Group CRC Press.

Mauluddin, S., Ikbal, I., & Nursikuwagus, A. (2020). "Complexity and performance comparison of genetic algorithm and ant colony for best solution timetable class," *Journal of Engineering Science and Technology,* 15(1):278–292.

Mukhairez, H. H., & Maghari, A. Y. A. (2015). "Performance comparison of simulated annealing, GA and ACO applied to TSP," *International Journal of Intelligent Computing Research (IJICR),* 15(1). http://hdl.handle.net/20.500.12358/25118

Islam, S., Tanzim, M., Afreen, S., & Rozario, G. (2019). "Evaluation of ant colony optimization algorithm compared to genetic algorithm, dynamic programming and branch and bound algorithm regarding travelling salesman problem," *Global Journal of Computer Science and Technology.* Retrieved from https://computerresearch.org/index.php/computer/article/view/1855

Şenaras, A. E., Sezen, H. K., & Şenaras, O. M. (2017). "Analyzing robot lines using discrete event simulation", The Journal of Academic Social Science, 39:435–446.

Şenaras, A. E., &, İnanç, Ş. (2018). "VBA implementation via dynamic programming for AGV line," *Journal of Life Economics,* 5(4):255–264.

Şenaras, A. E., & Şenaras, O. M. (2020). "Supply chain management via system dynamics in flexible manufacturing system." In Handbook of Research on Developments and Trends in Industrial and Materials Engineering (pp. 438–450). IGI Global.

Valdez, F., Moreno, F., & Melin, P. (2020). "A comparison of ACO, GA and SA for solving the TSP problem." In: Castillo O., Melin P. (eds), Hybrid Intelligent Systems in Control, Pattern Recognition and Medicine. Studies in Computational Intelligence, Vol 827. Springer.

Whitley, D. (1994). "A genetic algorithm tutorial," *Statistics and Computing,* 4(2):65–85.

# 5 Intelligent Control Design Schemes of a Two-Link Robotic Manipulator

*Ranjan Kumar*
Birla Institute of Technology, Mesra
Ranchi, India

*Kaushik Kumar*
Birla Institute of Technology, Mesra
Ranchi, India

**CONTENTS**



## 5.1 INTRODUCTION

Looking back in the historical perspectives, we come to know how the 'robotic system', perhaps more fittingly called 'human-like creature', has evolved and developed continuously as a result of the relentless efforts on research and utilization of multiple computational techniques during the last few decades. The term 'robotics' has a very significant meaning of being 'forced labourer' and was first appeared in the play 'Rossum's Universal Robots' (RUR) written by a Czech writer Karel

Čapek in the year 1921. In this play, the complete fictional mechanical creatures were designed so as they can work like a human being and intended to replace the human workers. The main idea was to show that these creatures are very much efficient but fully emotionless and are capable to perform the specified jobs according to the given instructions by maintaining the high level of accuracy and efficiency, and also proved to be very much useful in hazardous situations without any harm to human workers. With rapid technological growth, the robotics technology is also advancing itself with the pursuit of new ways in science. Nowadays, with the rapid economic and technology changes, the industries are also moving from the current manual state to automation and also towards the complete robotization. But along with this wide developmental arena, the feeling of fear and hatred also exists and has not yet lessen till date. The fear of occupying the people's job have reported some retardations in the application of the robotics in many areas. However, this takes us back to the year 1940 when Isaac Asimov's science-fiction story postulated robot as a 'human helper' [1-4].

After getting inspiration, Joseph F. Engelberger was very fascinated about developing a working robotic model and during 1950 to 1958 he started a company called 'Unimation' along with his fellow George C. Devol in the USA. Since then, the field of robotics evolved and found its multi-directional applications in areas such as painting, spraying, welding, inspection and loading and unloading of machine tools. Even some specific application involves micro-robots, underwater robots, path-tracing walking robot, surgery robots, etc. [5-6]. A robotic manipulator has to perform its specified task in a very unpredictable, hazardous and hostile environment such as space exploration, material transportation for space application and handling of dangerous radioactive materials. In space exploration, the robot has to be smart enough to coordinate and navigate [7-10] in unknown, unpracticable and unmodeled environment during space locomotion [11], inspections and satellite maintenance. Hence, for such applications the design of an intelligent control strategy is essential, and including smart and intelligent techniques such as adaptive control strategy is best suited because of the related issues in the inertial space which mainly comprises: (i) uncertainty and lack of joint trajectory and (ii) non-linear parametrization in inertial frame. The smart and intelligent system was very much accepted and admired among academicians, and the adoption rate among industries was appreciably high. But till now the intelligent system based on artificial intelligence (AI) and its applications were not so popular in manufacturing and among the robotics community in industries.

An intelligent robot may be defined as a completely equipped robotized mechanical system. The design of a robotic manipulator can be better understood with the help of a robot life cycle. The life cycle can be described in many phases of a robotic system comprising: design specification and simulation, system requirement, operational performance, operational maintenance, testing, assembling, dismounting and recycling of the system and their corresponding parts. All these steps discussed in design phase may differ as per the product or component difference. Also, the design and development may split into various design process such as: conceptual design, detailed design, visual design, parametric design, preliminary design, etc., as per the system requirement. With time, the multiple robotic system, such as path-tracking robot, obstacle avoidance, collision avoidance, climbing robot, jumping robot and robotics in tele-medicine, presents a diverse picture of its continuous growth and developments [12-13]. However, most of the literatures in this field of robotics talk about a wide range of smart intelligent system of AI-based design as an application of motion and path planning [14] that includes the smart Knowledge-based system (KBS), expert system (ES) and multi-agent system (MAS), along with the soft computing techniques such as fuzzy logic systems (FLS), genetic algorithm (GA), neural network (NN) systems and many more. Each of the above-mentioned systems may occur as stand-alone system or in combination with one another. 'There are two main categories of AI developments. The first includes methods and systems that simulate human experience and draw conclusions from a set of rules, such as expert systems. The second includes systems that model the way the brain works, for example, artificial neural networks'.

The availability of numerous literatures on each type of smart system produces a very diverse picture, along with their advantages and disadvantages. The main objective of AI [15] is to understand how a computer system can be made to show its intelligence and, in this process, we generally take the help of different soft computing algorithm and machine learning techniques to introduce the smartness or intelligence to a mechanical system such as a robotic manipulator. Earlier, the notion of intelligent system dealt directly with the robotic system as a preferred model of a smart and intelligent agent. But later on, during the 1970s the two fields got clearly separated and the field robotics became more focussed on the industrial automation, whereas the AI was focussed on utilizing these robots just to demonstrate the mechanical intelligence in day-to-day environment [16]. Indeed, this is quite difficult to summarize the borderline between the works done in the field of Robotics and Artificial Intelligence. However, the research community have already identified the problems associated in developing an intelligent robot and, these intelligent robots are viewed as a prototypical case [17] of the AI system. Further, by looking at the connections and relationship of 'AI-robotics', we come across two major associated issues of action and perception. In action, the basic structure of an autonomous agent, i.e. a robot system, is equipped with various sensing and actuating devices and the actions persists according to the data flow from sensors to actuators, which follows 'sense-plan-act' cycle repeatedly. This further includes an architecture of reaction which directly focuses on basic functionalities such as the interpretation of sensors and navigations of the robotics system. The reactive architecture generally addresses the environmental dynamics of the system. The logical view of this reactive functionality leads towards another field of 'cognitive robotics' which aims at developing an agent that can achieve a particular specified task with known dynamics in an unpredictable complex real-world environment without having any human assistance [18-21]. The robotic perception is a distinguished field in AI-robotics research. While interacting with the unmodeled, undefined and unstructured environments populated with many moving unknown object, robotic agent has to tackle and understand the environmental perception which involves high-level of descriptive perceived world.

Over the decades, many methods and computational approaches have been developed in order to design and control a smart and intelligent agent, i.e. robotic systems or robotic manipulators, using different control strategies based on the AI principles and machine learning algorithms. Some important and much widely used control schemes are: proportional-integration-derivative (PID) control [22], feedback control [23], fuzzy logic (FL)-based control [24], neural network (NN)-based control [25], adaptive control method [26], sliding mode control [27], observer-based control [28], LQR-based control [29] and genetic algorithm (GA)-based control [30]. The intensive literature survey highlights three main intelligent controller design techniques, viz. KBS, fuzzy system and neural network, for controller design even for the systems having unknown model and unknown dynamics.

In the present scenario, the world is very much fascinated and inclined towards the design of an intelligent model-based or intelligent model-free system with their known or unknown dynamics. The present work understands the requirement and aims to summarize three main controller design techniques: PID controller, fuzzy logic (FL)-based controller and the artificial neural network (ANN) based controller.

## 5.2 KINEMATICS AND TRANSFORMATION

A robotic manipulator has to perform certain jobs in a particular predefined path in the three-dimensional space. The end-effector is expected to follow that predefined particular trajectory to carry out the work in the workspace. Hence, it is necessary to have control over the position and joints of each link in order to obtain the desired trajectory and performance. A robotic system consists of the serially connected multiple links called 'joints', and the degree of freedom (DOF) of a robot depends on the number of associated links and joints, their types and the kinematic chain. Each individual part or the rigid body connected like a chain that makes up a complete robot are called links. For a rigid body, there occurs six DOFs in a three-dimensional cartesian space. If there is no relative motion between two or more links, then from the kinematic point of view of the system it is said to have a single link. The kinematics and mechanisms as well as the dynamic modelling of a system play a very vital role in designing a robotic manipulator. The kinematic modelling of a mechanical manipulator provides the necessary information related to the position and orientation of the end-effector as well as the spatial positions of the joint links in a three-dimensional space. Also, the differential kinematics of a robotic system derives all the higher order derivative of the position and describes the corresponding differential motion of the system such as velocity and acceleration. The derivative of the kinematic information deals with the 'kinematics of motion' and this does not bother about the forces or torques applied on the joints and basically deals with the dynamic and mathematical modelling of the system [31-34].

The number of DOFs of a manipulator is nothing but the number of independent parameters that are required to completely describe the position and orientation in the spatial frame of reference [35]. Reuleaux is known to be the 'father of kinematics' for his enormous contribution in kinematics and theory of machines. The key idea of his kinematic theory was that he recognized the mechanisms of a machine as a chain of constrained elements and by looking at this chain he understood the existence of the geometrical constraints between the kinematic pairs. The number of variables required to provide the complete description of the system, Reuleaux [36] defined some important symbolic notations. His idea of proposed symbolic notations provides the kinematic description, kinematic property and the analysis of the system, which became the base for the synthesis of new mechanisms of a system. But the symbolic notations proposed by him was not enough to provide the complete kinematical description of the system. Denavit and Hartenberg [37] reported this lacking of symbolism and proposed the reconsideration for the kinematic symbolic notation, which completely defines kinematic properties using the lower pair by means of kinematical equations. The study of symbolic notation is based on the kinematical properties and it uses the properties of matrix algebra provided by solved examples. In machine parts, when the motion of each relative part is constrained or guided in a unique fashion, then the main purpose of mechanisms is to provide the transformation of motion from one part to another. This transformation of motion can be best seen in the slider-crank mechanism where the linear motion of piston-cylinder arrangement transformed into the continuous circular motion. Here, the connections between the adjacent parts are used to produce the relative motions by means of the existing surface contacts between the parts. Therefore, the main objective for designing such mechanisms is to establish the connection between the relative pairs in such a proportion so as to produce the desired output motion corresponding to an input motion. The two parts together are known as 'pair' and each individual part of the pair is termed as 'element of the pair'. The existence of the surface contacts between the elements of the pair is termed as 'lower pair', and when these contacts between the elements of the pair is confined to a line, it is termed 'higher pair'. For the sake of their relative simplicity, the lower pair of a mechanism is studied first. The possible lower pair can be: spherical pair, cylindrical pair, plain pair, screw pair, revolute pair and sliding pair.

Whittaker [38] has reported about many of the kinematical theories and geometrical proofs along with many other theorems provided by Euler and Chasles and has traced their work. The central idea of the study is to provide a big picture on the successive positions of a rigid body in a three-dimensional space by actually providing the linear transformations or sometimes by providing both translational and rotational transformation. The study of finite displacement of a rigid body is essential for obtaining the pose of a rigid body and by means of computation, the solution to the transformations of rigid body in 3D space can be estimated. To find the relative positions and orientations of a rigid body from rotational and translational parameters, the work of Paul [39], Suh and Radcliffe [40] and Beggs [41] can be accessed and worth noted. Laub and Shiflett [42] came up with an idea that the position and orientation of a rigid body can be estimated in a three-dimensional space just by defining three fixed non-collinear points in the rigid body. By a series of composite transformations (translation and rotation), the motion of a rigid body can be estimated and the sequence of displacements can be formulated in a matrix form. The provided approach can describe the pose of a rigid body even in a simpler manner from the imprecise data obtained from the measurement of the positions of the three non-collinear points.

## 5.3 TWO-LINK MANIPULATOR DYNAMICS

It is known that a robotic manipulator has to perform a specified task in a three-dimensional space. In order to perform the tasks such as the lifting of workpiece from a machine, the end-effector moves in the three-dimensional space in a pre-defined path. In this process, the end-effector exerts some forces and moments to the environment at the contact points. These forces and moments are developed by the pre-installed actuators at multiple joints of the manipulator. These developed actuator forces and moments are further transmitted to the contact points through the open chain [42-43]. In the statics of robot manipulator, the relationships between the joint torques, joint forces, the moments in cartesian space and the forces exerted on the end-effector are estimated. The motion in the robotic system arises due to generated torques/forces at the joints. For describing the dynamical behaviour of the manipulator, it is necessarily important to develop a dynamic model of a particular system. This mathematical modelling of the manipulator is very helpful in describing many things: (i) dynamical model is necessary to design the sophisticated control system to achieve the optimal performance of the said manipulator, (ii) this dynamical model can be used to estimate the real-world performance of the manipulator using the computer simulation, (iii) the joint reaction forces and moments require the dynamic analysis for estimating the link sizes, bearings and actuators for the robotic manipulator and (iv) some of the controllers depend on the dynamical model to compute the actuator torques and forces for desired trajectory tracking. Hence, for developing the dynamical model, there are number of methods that can be availed.

The mathematical formulation of these methods can be given by Lagrange-Euler formulation [44-49], Newton-Euler dynamic approach [50-51], Kane's dynamical method [52-53] and recursive Lagrange method [54-56]. Out of all these proposed methods, the Newton-Euler dynamic approach and the Lagrange-Euler formulations are very much efficient and are much widely used in manipulator modelling. These two formulations possess their own advantages and disadvantages.

The efficient dynamic modelling is important especially for the advanced control and manipulator design. Thomas and Tesar [48] derived the dynamic model for the serial rigid manipulator. The proposed dynamics reduces all the dynamic properties in the generalized inputs. Also, the components of the dynamics had been calculated using the dynamic influence coefficients and presented in a very lucid form. The efficient dynamic modelling with extensive description can be found in Tesar and Thomas [57]. By knowing the physical parameters, the manipulator dynamics problems can be solved using: (i) inverse dynamics and (ii) forward (direct) dynamics. The inverse dynamics proposed by Kane and Levinson [52] is used to calculate the joint torques or joint forces produced by actuators and is required to generate the trajectory for a robotic system. It is useful in computer simulation to show how a robot is going to respond in the real world when it is built. The forward dynamics is necessary for obtaining the responses for a robot. It actually estimates the performance on the basis of the torques/forces applied at the joints. Further, a new form of Lagrange formulation has been reported by Chang [58], which provides a well-defined equation of motion in the form of second-order differential equation and shows highly non-linear and coupled form of equation and is much useful for both controller design

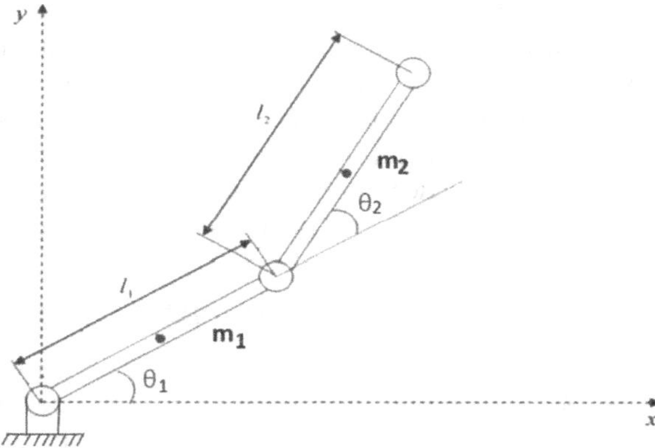


**FIGURE 5.1** A basic schematic diagram of a two-link robotic manipulator.

and joint forces/torques estimation. The robot manipulator shows highly non-linear, uncertain and coupled dynamics with time-varying parameters, which are mostly used in industrial applications [59-62]. Hence, it is important to control the manipulator because of the associated uncertainties and non-linear dynamics in a way that it could follow the desired trajectory and can produce the required performance. The schematic diagram of a two-link manipulator is given in Figure 5.1.

The complexities involved in a human arm [63] can be studied by modelling a two-link robotic manipulator as shown in Figure 5.1. Developing a dynamic model for a manipulator design leads us towards the dynamic behaviour and allows us to understand the system's characteristics in real-world scenario. The dynamical effects such as the inertia force, Coriolis force and centrifugal force produced by the joint actuators also influence the manipulator's performance. The joints of the manipulators got activated through the power transmission provided by the attached motors, gears or belt [64]. The Lagrange formulation of $n$-links robotic manipulator connected serially can be described by the equation of motion as:

$$M(\theta)\ddot{\theta}+V(\theta,\dot{\theta})\dot{\theta}+G(\theta)=\tau \tag{1}$$

where $\theta$ is the joint variable vector showing the relative positions, $M(\theta)$ is the non-singular inertia matrix, $V(\theta, \dot{\theta})$ gives the collecting centrifugal and Coriolis force/torque, $G(\theta)$ is the vector of conservative forces and $\tau$ is the vector of generalized forces produced by motors.

## 5.4 CONVENTIONAL CONTROL

Since many years, the 'conventional controllers' such as PID and computed control torque methods have been used in the many industrial applications. The traditional controllers have simple control, ease of design, low costs and convenient implementations. The three modes of this traditional PID controller, i.e. proportional (P), integral (I) and derivative (D), make any system 'robust' and 'efficient' to provide the desired outputs corresponding to the input singles. These outputs are produced in terms of the 'rise time, peak time, overshoot, settling-time and steady-state error'. These conventional control methodologies are also associated with some problems: (i) requirement of mathematical model (or system dynamics), (ii) cannot sustain with variational payloads and (iii) sudden change of the plant parameters due to uncertainties results inaccurate performance of the controller.

### 5.4.1 Proportional-Integration-Derivative (PID) Control

It is a well-known fact the PID [65] controller is one of the extensively used control tools in industrial applications and automation sector. It is estimated that over 90% of the control loop is based on the PID control. The basic schematic diagram of a PID controller is given in Figure 5.2.

The basic control law which is most commonly used in PID control structure can be defined mathematically as follows:

$$\tau_{PID} = K_P e(t) + K_I \int e(t)dt + K_D \frac{de(t)}{dt} \tag{2}$$

The three important terms [66] included here in the basic PID control structure are proportional, integral and derivative, and $K_P$, $K_I$ and $K_D$ are the corresponding controller gain matrices. Also, $\tau_{PID}$ is termed as the vector of joint torques or joint forces. All these three terms of a basic PID controller are very much essential to be applied in the most common control problems. The integral control is very helpful in tracking the constant setpoints and the result is described by the internal model principle and is demonstrated by using the final value theorem. The integral control also enables the disturbance rejection and it is also helpful in the noise filtration produced by the higher frequency sensors but it also responds slow to the current error. The proportional control responds immediately to the current error but the high level of accuracy in setpoints is quite difficult to obtain without having the unexpected large gain values. The derivative term amplifies the higher frequency sensor noise. These three modes of PID controller make any system much efficient to produce the desired output in terms of rise time, settling time, overshoot and steady-state error [67-69]. Manjeet and Khatri. [70] have reported an efficient controller for a robot manipulator which ensures the desired trajectory-tracing and performance output.

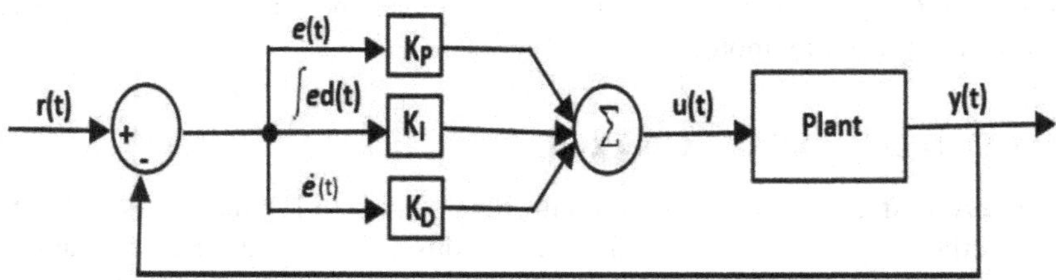


**FIGURE 5.2** Basic schematic diagram of a PID control structure.

The most common type of controllers used in industries is PID controller, which produce better performance, fast time-response and low steady-state error [71]. The end-effector's defined position and defined speed are considered as input parameters and the desired position and desired speed are treated as output parameters. Further, in this sequence Shah et al. [72] have designed a hybrid controller which is a combination of both traditional 'PID-type' and 'computed torque control' (CTC)-type controller, and the simulated result gives a better performance. This is basically a 'mass-only' system which is a very simplified form of system designed to obtain the velocity and acceleration error through the simulation process.

### 5.4.2 Computed Torque Controller

As we have discussed, a robotics manipulator is a series of connection of rigid links with one end fixed and another end free. The whole body of a manipulator is joined together by revolute or prismatic joints. The dynamic equation of motion of a manipulator is given in equation 1. The control law of the CTC method is expressed as:

$$\tau = \hat{M}(q)u + \hat{V}(q,\dot{q}) \tag{3}$$

$$u = \ddot{\theta}_d + K_P(\theta_d - \theta) + K_V(\dot{\theta}_d - \dot{\theta}) \tag{4}$$

where $\hat{M}$ and $\hat{V}$ are the estimations of $M$ and $V$, respectively, used for non-linear compensation. Also, $K_P$ and $K_V$ are proportional and derivative gain values of the controller. The CTC method [73] fulfils the efficient 'motion control objective' and is used to obtain a desired trajectory performance of a robot manipulator effectively. The CTC was one of the first model-based 'motion control strategies' developed for manipulators. The computed-control method (CTC) ensures the global asymptotic stability for fixed symmetric positive-definite (proportional and derivative) gain matrices. As the global 'asymptotic stability' ensured in the computed control method is for 'fixed symmetric positive-definite' (proportional and derivative) gain matrices. The CTC strategy requires an accurate system dynamic for the purpose of desired performance and accuracy. Song et al. [71] have designed fuzzy-based computed torque controller for trajectory tracking of robotic manipulator in the very certain or uncertain/unpredicted environments in structured or unstructured manner. The said controller provides stability guarantee based on the 'Lyapunov's stability theorem'. The CTC strategy is quite simple and easy to understand, and uses the 'feedback linearization techniques' to give a 'linear time-invariant closed-loop system'. Hence, this controller is also termed as 'inverse dynamics control' [72-73].

## 5.5 INTELLIGENT CONTROL SCHEMES

The ability of sensing, understanding and learning based on the process characteristics, disturbance characteristics and operational practices is defined as 'intelligent control'. In the context of our field of intelligent controller, intelligent means strong man-machine interactions and associations [74]. This man-machine interaction in a very unpredictable environment is supposed to move one step further where it requires advanced decision-making [75]. The intelligent control techniques or strategies have paved its own way and have opened a door towards the spectrum of complex design practice and applications. The associated complexities in the system involve a wide range of characteristics, including the 'poor model, decision space, high dimensionality, distributed sensors, systems with multiple subsystems, high noise level, time scales or performance criteria, complex information pattern', etc. For a living being, it is quite easy to get adapted to these complexities in real life, but it is very difficult for machines to adopt those characteristics. In the last few decades, many theorists are believed to have successfully designed an intelligent system by putting their enormous efforts on introducing those many characteristics to an artificial system in an artificial way. Now, in the present, as a result many of the phycological and biological terms and techniques have been adopted, such as pattern recognition, adaptation, self-organization and learning capabilities, and have been introduced into the literatures [76]. It is important to develop a mathematical model for a system or process which is to be controlled. But sometimes it is possible to have lack of system dynamics or poor system dynamics or unknown model for which a controller is to be designed. Here, the intelligent control strategy comes into play where we require some linguistic information which should be purely based on the human experiences and expertise for designing a model-free controller for that particular system. In this context, the literature suggests some intelligent controllers, namely, 'fuzzy logic control (FLC), ANN control and neuro-fuzzy controller, genetic algorithm (GA)', which are used mostly towards the intelligent control design [77].

### 5.5.1 Fuzzy Logic Control

Generally, it is better to have a mathematical model in order to design a controller for a particular system for the ease of design, and from the accuracy point of view, the mathematical model for a system, or, in our case, manipulator, helps us to obtain the desired system performance based on the designed and applied controller logics. But, sometimes, we come across situations where we even do not have the proper knowledge of the system or system specifications, and even the lack of the corresponding dynamics of the system leads us in the pursuit of some other design technique for implementing the intelligence to the machine. Therefore, to fulfil such requirements a model-free fuzzy logic (FL) control design technique was introduced to obtain the approximate and inexact behaviour corresponding to the environmental changes. For such 'unknown modelled system', the fuzzy logic-based control design approach is an algorithm that requires some 'linguistic terms' based on the human expertise and experiences which connects linguistic control strategy and specific control strategy together [78-80]. Generally, an FLC contains mainly four inbuilt components: (i) fuzzy rule base, (ii) fuzzy inference engine (this is the decision-making logic), (iii) fuzzification interface module and (iv) defuzzification interface module. Out of these four components, the fuzzy inference engine is the kernel of the fuzzy logic controller [81]. A schematic view of a FLC is shown in Figure 5.3.

The main purpose of applying the fuzzy logic is that the FLC provides a standard approach for easy adaptation and easy application of FLC rules from the rule base of

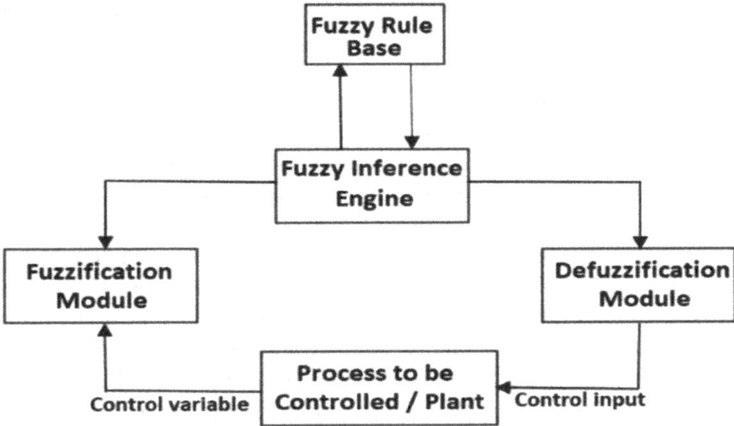


**FIGURE 5.3** A schematic view of a fuzzy logic control.

the fuzzy logic system. Also, this helps us to overcome the problem of using the heuristics approaches. This can be achieved by modelling the non-linear parameters of a robotic manipulator using a set of FLC IF...THEN rules for the unknown parameters. A complete simulation result of a two-link robotic manipulator using the FLC scheme can be found in Lai and Shieh [82]. The simulation results show the position error of the manipulator on following a cubic-spline trajectory; also, by changing the trajectory from the cubic-spline path to the rhombus path, the robustness of the controller is verified.

The performance of an FLC depends on its knowledge base (KB) approach, which consists of (i) data base (DB) and (ii) rule base (RB). However, the proper KB designing of an FLC is difficult, which can be implemented by using one of the following four ways: (i) optimization of DB only, (ii) optimization of RB only, (iii) stage-wise optimization of DB and RB and (iv) simultaneous optimization of both DB and RB. It is advisable to design minimum rule base for an FLC in order to have easy implementation (either in software or hardware) of the controller and the less computational complexities because a large number of rule base require a large number of linguistic information, resulting in uncontrolled computational complexities [83-84]. Realizing these facts and requirements, the development of a hierarchical FLC has been investigated [85-86] with only one objective, i.e. to keep the number of rules as minimum as possible.

It is a well-known fact that, for a non-linear robotic manipulator and their associated uncertain parameters, the FLC type controller is pretty much efficient and the most widely used tool in multiple applications in many research fields. In today's scenario, the robotic manipulators have to work efficiently in a very unpredictable, unmodeled and uncertain environments. Hence, with continuous technological change and advancements according to the current problems and needs, the researchers are much focussed and are inclined towards developing the hybrid controllers (Fuzzy P+ID) [87-88]. In this context, Anavatti et al. [89] have reported PD-type fuzzy controller in combination with the conventional PID-type controller.

The demonstrated solution has proved the better efficiency and appropriate solutions along with Lyapunov's stability [90] method towards controlling the manipulator. The tuning of the optimal parameters can be done online or offline to obtain the accurate and desired performance of a manipulator just by minimizing the errors and deviations. Piltan et al. [91] have carried out a study based on a tunable gain of a PID fuzzy controller, termed as 'GTFLC', for a three-DOF robotic manipulator design which have been formulated by a strong mathematical background just to develop an 'adaptive non-linear robust' controller with acceptable performances producing the error minimization, better disturbance rejection and desired trajectory-tracing capability. The FLC-based controller has larger application area, but the calculations and tuning the PID system are very tedious job. Hence, the main advantage of using an online tunable gain is that the 'combined online tunable gain' can solve the associated uncertainties from the uncertain non-linear system [92-94].

The conventional PID-type controller is associated with many flaws such as: (i) the 'computation of errors', (ii) 'noise reduction in derivative control', (iii) losses in performance and oversimplification and (iv) complication occurring due to integral control law. Corresponding to these main shortcomings, Han [95] has proposed four unique measures: (i) differential equation: used for 'transient trajectory generation', (ii) 'a noise-tracking tolerant differentiator', (iii) 'a non-linear law' and (iv) methods and concepts for estimating the rejection rate of disturbances and noises. Hence, to enhance the control performance due to various flaws in PID, Podlubny [96] has investigated and introduced a new concept of 'fractional order' or 'arbitrary real order' control and $PI^{\lambda}D^{\mu}$ controller. The fractional-order calculus [97] is a separate branch of mathematics in which the fundamental fractional-order operator is used for generalization of differentiation and integration of the five important parameters, viz. $K_P$, $\mu$, $K_D$, $\lambda$, $K_I$, of this fractional-order PID, i.e. FOPID controller which basically needs a global optimizer such as genetic algorithm (GA) for automatic optimization. Further, for designing a hybrid system, it is important to transform the existing system, i.e. traditional PID-type controller, into an equivalent FLC-type controller. Realizing this fact, Chao et al. [98] have investigated on the optimal fuzzy controller [99] for a system. The core idea of this approach provided the transformations of a traditional PID control to an equivalent FLC within the ranges defined for the input-output values [100]. Also, in this process of transformation, the obtained values of membership functions (MFs) get optimized by tuning according to the specified non-linear factors of the system. The tuning of the parameters for any specified system is much important for controlling purpose and this tuning is totally based on the experiences and human expertise; also, it is well defined under the guidance of some principles suggested by Ziegler and Nichols [101].

Following the desired defined trajectory, accuracy and robustness of the system can be included as some major advantages of the FLC [102-103]. FLC can be considered as a potential tool which can deal with the uncertainties and imprecisions involved in the system. This does not require much extensive mathematical formulation and also the rule base of FLC system follow IF...THEN form. This makes the user understand the control system in a much easier and efficient manner. The extensive applications related to this FLC have been demonstrated and can be found in Lim and Hiyama [104].

### 5.5.2 Artificial Neural Network

One of the intelligent methods, ANN comes with its own importance when we come across some unknown model or unknown system having unknown dynamics or mathematics. When we do not have a proper and accurate model, it is quite difficult to design a controller for that particular system for the purpose of controlling. The ANN is based on the principle of biological neurons or we can say it is inspired biologically [83]. The ANN is pretty much similar to our human brain, so this can be considered as a highly complex structure of a parallel computer. An ANN has a complex architecture of having a large number of neurons arranged in multiple layers and the neurons of each layer are connected to the neurons of other layer by means of weights, as shown in Figure 5.4.

Sometimes, we come across the unmodelled dynamics of a system, and corresponding to this system, the controller design is a very tedious job. Using the concepts of ANN here is proved to be well-suited and quite efficient. The ANN adds up the knowledge through the recognition of pattern among the large number of available datasets. On the basis of these datasets, the neural networks got trained by learning through experiences. The behaviour of an ANN model can be estimated by the associated transfer function, learning rule and its model architecture. The neural network introduces non-linearity to the system and it gets activated by an 'activation signal' passing through the transfer function and as a result it produces the output [105]. The structure of an ANN model can be considered as a 'computational parallel structure' which has been encouraged by neuroscience and the further evolution in this field can be studied in Anderson and Rosenfeld [106]. 'System identification' and 'control design' are two main stages of the ANN-based controller [107]. The first stage comprises the development of 'neural model', and after that, in the second stage, this 'neural model' is used to train the networks by continuous learning, reaching the optimal learning. When the network gets fully trained, one can obtain the output corresponding to any new input parameters.

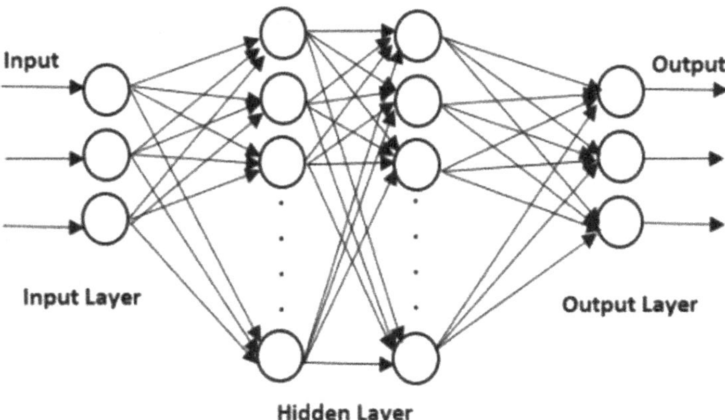


**FIGURE 5.4** The basic schematic diagram of a multi-layer neural network.

The feedforward control and the feedback control along with the filters can be implemented as important parts of an ANN model [108]. These two types of control can be combined together to solve the problems related to the 'classical control' and 'robust control'. In this sequence, Lewis et al [109] have proposed a design of a controller based on the 'multilayer neural network'. In this paper, the optimal performance of the system covering the desired trajectory-tracing control has been investigated. Also, the weights of the neural net are not needed to be trained online. So, the training and tuning of the system has been done offline which guarantees the systems performance; also, the stability of the path tracking has been investigated and the error convergence is found to be guaranteed [110]. Due to having high non-linearity and uncertainty involved in a robotic system, the controlling of the associated kinematics and dynamics of such systems is not so easy. Complications are involved for such systems in the form of computational inertia forces, joint reactions, joint forces or toques and joint friction. Hence, the occurrence of the intelligent systems or techniques over the conventional controller is very much essential but it involves some complications as mentioned above. Thus, the attainment of a meaningful model and establishing the relationships between these complications, the measured input-output data are pretty much essential and are much appealing [111-113]. To handle such situations Prabhu and Garg [114] have presented an overview on the applications of ANN in the field of robotic control and different learning methods. The paper also investigated some important related issues and strengths and weaknesses in the particular area.

## 5.6 RESULT

The present work has gone through an extensive literature survey on the different intelligent and knowledge-based control design for a robotic manipulator (Figure 5.5). Here, we have mainly discussed two important intelligent control techniques: the FLC and the ANN control. Their efficiency and performance over the traditional

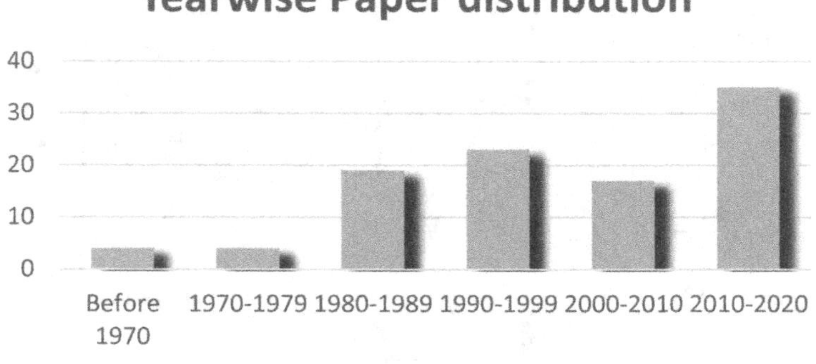


**FIGURE 5.5** Diagram showing the year-wise published paper distribution.

controllers such as PID and CTC method have been studied. The advantages and disadvantages for these two control schemes have also been highlighted. The main objective of a fuzzy logic approach is to develop an easy and well-structured technique for applying the concepts of FLC rules from the rule base and to make it possible to overcome the dependencies of using the heuristics methods. The fuzzy logic approach uses the linguistic information based on human expertise, which means that under the FLC scheme, the mathematical modelling of a system is not of much importance. On the basis of the literature review, it has been concluded that the FLC scheme suffers from the lack of 'self-tuning' capability and the 'real-time control' along with a flaw of having the requirements of 'large memory allocation'.

The ANN has well-defined and structured capability of continuous learning and 'learning by training' under uncertainties and for unmodeled unknown dynamics of the system. The ANN possesses 'small trajectory error and fast error convergence' and is invariant to the variational payload. All these occur during the training of the network, and when the network gets fully trained, the ANN model is now ready to put any input parameter for their corresponding desired output. This approach, being model-free, is easier for the user even for systems having unknown dynamics. However, since an ANN model gets trained by continuous learning and experience, with the process being iterative, the networks take large number of iterations to be trained fully. Therefore, this increases the computational time drastically, which is much annoying, and consequently the computational costs are high. In an ANN model, no well-defined approach is available to determine the number of neurons needed for training the networks. So, basically, we do not know how much time it is going to take for a particular task and for task with large associated variations. Also, the learning of the network does not guarantee error convergence.

## 5.7 CONCLUSION

The literature survey is sufficient to conclude this chapter which discusses the design and different control schemes of a robotic manipulator. In this chapter, the main focus has been kept on the intelligent controller. In this regard, two very popular and known soft computing approaches, viz. FLC and ANN, have been discussed, which provides us a better perspective towards the easy controller design over the mostly used conventional controller. These two soft computing techniques are associated with many common factors such as 'state space and distribution on the parallel processors'. The literatures highlight that a robotic manipulator shows highly non-linear and strong coupling; therefore, the classical traditional controllers are not that much efficient and robust to withstand 'high variational parameters, high variational payloads, uncertainties', etc., but still these traditional controllers are being used in large scale for industrial applications and for performing slow-speed tasks such as pick-and-place and others. It has also been shown in the literature that the discussed intelligent controllers, i.e. FLC and ANN, are much efficient and robust soft computing techniques which produce the desired performance of a manipulator in terms of 'trajectory tracking', 'error convergence' and 'quick response time'. These techniques are very much helpful in designing the smart intelligent controller for a robotic manipulator.

## 5.8 FUTURE SCOPES

In the very early stage of developments, the researchers were relying only on the AI algorithmic approach and they were trying to develop some expert knowledge from a specific domain. As time goes on, the continuous advancements have led us to think multi-dimensionally as per the requirements and the problem statements we have. We have now developed a strong foundation over the soft computing techniques, and due to having the high-speed computers, the computational complexities are not limited now. As far as the 'intelligent control' is concerned, today we are developing very powerful intelligent control techniques or methods which have characteristics such as 'quick adaptation, self-learning, self-organizing, self-diagnosis (includes reconfigurations and self-repair)', etc., of the system. The control system has been evolved from the 'conventional control to the intelligent expert control' through advanced intelligent computing techniques and still has to go further in the future to improve sophistication and robustness to minimize the flaws associated with today's techniques. Indeed, during the last few decades, the 'intelligent control' has been enormously developed but during the same time the 'powerful electronics hardware' has also been evolved parallelly, which has a strong practical implementation to predict the performance and intelligence. Further developments of more sophisticated intelligent robotic manipulator will be based on the 'smart sensors' and 'smart actuators' with more computational power, and the study says this will be possible as a result of current explorations in semiconductor chip technology. In the context of 'intelligent control', the FLC has been introduced in a wide range of application areas. Although it has gained much popularity, it still has some limitations, which is the scope of further research. Anything that reduces the product cost is valuable for customer and FLC plays this role very well and provide a model-free approach where no complex mathematical formulation is required but there is a need for evolution in its real-time performance and self-tuning capabilities such as NN control strategy. Also, the 'large-memory requirements' of FLC may be reduced in the coming future advancements. Similarly, an ANN model has many advantages but the large computation time due to multiple learning for getting trained networks makes it difficult for user, especially in problems involving many compilations and variations. In the coming future, the advancements have wide dimensions to work over these flaws and can achieve to have a more suitable and sophisticated robust controller for more complicated structures and dynamics or for unknown model and unknown dynamics.

## REFERENCES

1. R.C. Dorf, International Encyclopaedia of Robotics, John Wiley & Sons, New York, NY, 1988.
2. RobotShop Distribution Inc., "History of robotics: timeline," 2008. [Online]. Available at: http://www.robotshop.com/media/files/PDF/timeline.pdf
3. R.K. Mittal and I.G. Nagrath, Robotics and Control, McGraw Hill Education (India) Pvt Ltd, India, 2019.
4. S.K. Saha, Introduction to Robotics, McGraw Hill Education (India) Pvt Ltd, India, 2013.

5. Yangsheng Xu, "Parameterization Control of Space and Adaptive Robot Systems," *IEEE Transaction on Aerospace and Electronic Systems,* Vol. 30, No. 2, April 1994, pp. 435–451.
6. W.L. Whittaker and T. Kanade, Space Robotics in Japan, Loyola College, Baltimore, MD, 1991.
7. M. Ullman and R. Cannon, "Experiments in Global Navigation and Control of Free-Flying Space Robots," in Proceedings of the ASME Winter Conference, San Francisco, CA, 1989.
8. M.W. Walker and L.B. Wee, "Adaptive Cartesian Coordinate Control of Space Based Robot Manipulators," in Proceedings of the American Control Conference, Boston, MA, 1991.
9. M.W. Walker and L.B. Wee, "An Adaptive Control Strategy for Space Based Robot Manipulators," in Proceedings of the IEEE Conference on Robotics and Automation, Sacramento, CA, 1991.
10. Y. Xu and T. Kanade, Space Robotics: Dynamics and Control, Kluwer Academic Publishers, Boston, MA, 1992.
11. H. Ueno, Y. Xu and et. Al, "On Control and Planning of a Space Station Robot Walker," in Proceedings of the IEEE Conference on System Engineering, Pittsburgh, PA, 1990.
12. Shweta Patil and Sanjay Lakshminarayan, "Position Control of Pick and Place Robotic Arm," International Conference on Engineering Innovation and Technology, Nagpur, 2012.
13. Tian Huang and Songtao Liua, "Optimal Design of a 2-DOF Pick-and-Place Parallel Robot Using Dynamic Performance Indices and Angular Constraints," *Mechanism and Machine Theory,* Vol. 70, 2013, pp. 246–253.
14. George L. Kovacs, "Some artificial intelligent techniques to design robotic systems," in: L.M. Camarinha-Matos, H. Afsarmanesh and V. Marik (eds), Intelligent Systems for Manufacturing, Springer, Boston, MA, 1998, pp. 265–278.
15. Michael Brady, "Artificial Intelligence and Robotics," in: Michael Brady, Lester Gerhardt and H.F. Davidson (eds.), Robotics and Artificial Intelligence, Springer-Verlag, Berlin Heidelberg, 1984.
16. Antonio Chella, Luca Iocchi, Irene Macaluso and Daniele Nardi, "Artificial Intelligence and Robotics," *Intelligenza Artificiale,* Vol. 9, 2006, pp. 87–93.
17. S.J. Russell and P. Norvig, Artificial Intelligence: A Modern Approach, Pearson Education, New York, 2003.
18. Giuseppe De Gaicomo, Luca Iocchi, Daniele Nardi and Riccardo Rosati, "A Theory and Implementation of Cognitive Mobile Robotics," *Journal of Logic and Computation,* Vol. 9, No. 5, 1999, pp. 759–785.
19. Hans-Hellmut Nagel, "Steps toward a Cognitive Vision System," *AI Magazine,* Vol. 25, No. 2, 2004.
20. P.C.R. Lane and F. Gobet, "Simple Environments Fail as Illustrations of Intelligence: A Review of R. Pfeifer and C. Scheier, Understanding Intelligence," *Artificial Intelligence,* Vol. 127, No. 2, 2001, pp. 261–267.
21. A. Chella and M. Frixione, "Understanding Dynamic Scenes," *Artificial Intelligence,* Vol. 123, 2000, pp. 89–132.
22. J.T. Wen and S.H. Murphy, "PID Controller for Robot Manipulator," CIRSSE Document #54, Rensselaer Polytechnic Institute, 1990.
23. C.C. Cheah, S. Kawamura and S. Arimoto, "Feedback Control for Robotic Manipulator with Uncertain Kinematics and Dynamics," International Conference on Robotics & Automation, Leuven, Belgium, May 1998, pp. 3607–3612.
24. S.M. Prabhu and D.P. Garg, "Design of a Fuzzy Logic Based Robotic Admittance Controller," *Intelligent Automation and Soft Computing,* Vol. 4, No. 2, 1998, pp. 175–190.

25. M.J. Willis, G.A. Montague, C.D. Massimo, M.T. Tham and A.J. Morris, "Artificial Neural Networks in Process Estimation and Control," *Automatica,* Vol. 28, No. 6, November 1992, pp. 1181–1187.
26. P. Tomei, "Adaptive PD Controller for Robot Manipulators," *IEEE Transactions on Robotics and Automation,* Vol. 7, 1991, pp. 565–570.
27. Florin Moldoveanu, "Sliding Mode Controller Design for Robot Manipulators," *Bulletin of the Transilvania University of Braşov,* Vol. 7, No. 2, 2014.
28. Wen-Hua Chen, "Disturbance Observer Based Control for Nonlinear Systems," IEEE/ASME Transactions on Mechatronics, Vol. 9, No. 4, December 2004, pp. 706–710.
29. J.H.C. Rojas, R. Rodriguez, J.A.Q. Lopez and K.L.R. Perdomo, "LQR Hybrid Approach Control of a Robotic Arm Two Degrees of Freedom," *International Journal of Applied Engineering Research,* Vol. 11, No. 17, 2016, pp. 9221–9228.
30. K. Krishnakumar and David E. Goldberg, "Control System Optimization Using Genetic Algorithms," *Journal of Guidance, Control, and Dynamics,* Vol. 15, No. 3, May-June 1992, pp. 735–739.
31. J.J. Craig, Introduction to Robotics: Mechanics and Control, Addison-Wesley, Singapore, 1986.
32. D.K. Pratihar, Fundamental of Robotics, Narosa Publishing House Pvt Ltd, New Delhi, 2019.
33. C.S.G. Lee, "Robot Arm Kinematics, Dynamics and Control," *IEEE Computer,* Vol. 15, No. 12, 1982, pp. 62–80.
34. J.Y.S. Luh, "An Anatomy of Industrial Robots and Their Controls," *IEEE Transactions on Automatic Control,* Vol. 28, No. 2, February 1983, pp. 133–153.
35. Francis C. Moon, "Franz Reuleaux: Contributions to 19$^{th}$ Century Kinematics and Theory of Machines," Applied Mechanics Reviews, Vol. 56, No. 2, 2003, pp. 261–285.
36. F. Reuleaux, Kinematics of Machinery, Translated and edited by A.B.W. Kennedy, Macmillan and Company Ltd, London, 1876.
37. J. Denavit and R.S. Hartenberg, "A Kinematic Notation for Lower-Pair Mechanisms Based on Matrices," *ASME E, Journal of Applied Mechanics,* Vol. 22, 1955, pp. 215–221.
38. E.T. Whittaker, A treatise on Analytical Dynamics of Particles and Rigid Bodies, Cambridge University Press, Cambridge, MA, 1927, Chapter 1.
39. B. Paul, "On the Composition of Finite Rotations," *American Mathematical Monthly,* Vol. 70, 1963, pp. 859–862.
40. C. Suh and C. Radcliffe, Kinematics and Mechanisms Design, Wiley, New York, 1978, Chapter 3.
41. J.S. Beggs, Advanced Mechanism, Macmillan, New York, NY, 1966, Chapter 2.
42. A.J. Laub and G.R. Shiflett, "A Linear Algebra Approach to the Analysis of Rigid Body Displacement from Initial and Final Position Data," *Journal of Applied Mechanics,* Vol. 49, March 1982, pp. 213–216.
43. J.Y.S. Luh, "Conventional Controller Design for Industrial Robots — A Tutorial," *IEEE Transactions on Systems, Man, and Cybernetics,* Vol. 13, No. 3, May-June 1983, pp. 298–316.
44. R.H. Critchlow, Introduction to Robotics, McMillan Publishing Company, New York, 1985.
45. R.P. Paul, Robot Manipulators: Mathematics, Programming, and Control, The MIT Press, Cambridge, MA, 1981.
46. A.K. Bejczy, "Robot Arm Dynamics and Control, JPL-TM 33-669, February 1974.
47. S. Megahed and M. Renaud, "Minimization of the Computation Time Necessary for the Dynamic Control of Robot Manipulators," Proceedings of the 12th International Symposium on Industrial Robots, Paris, June 1982, pp. 469–478.

48. M. Thomas and D. Tesar, "Dynamic Modelling of Serial Manipulator Arms," *ASME Journal of Dynamic Systems, Measurement, and Control,* Vol. 104, 1982, pp. 218–228.
49. M. Vukobratovic, Shi-Gang Li and N. Kircanski, "An Efficient Procedure for Generating Dynamic Manipulator Models," *Robotica,* Vol. 3, 1985, pp. 147–152.
50. J.Y.S. Luh, M.W. Walker and R.P. Paul, "On-Line Computational Scheme for Mechanical Manipulators," *ASME Journal of Dynamic Systems, Measurement, and Control,* Vol. 102, 1980, pp. 103–110.
51. Y. Stepanenko and M. Vukobratovic, "Dynamics of Articulated Open-Chain Active Mechanisms," *Mathematical Biosciences,* Vol. 28, 1976, pp. 137–170.
52. T.R. Kane and D.A. Levinson, "The Use of Kane's Dynamical Equations in Robotics," *The International Journal of Robotics Research,* Vol. 2, No. 3, 1983, pp. 1–21.
53. R.L. Huston, C.E. Passerello and M.W. Harlow, "Dynamics of Multirigid-Body Systems," *ASME Journal of Applied Mechanics,* Vol. 45, No. 4, 1978, pp. 889–894.
54. J.M. Hollerbach, "A Recursive Lagrangian Formulation of Manipulator Dynamics and a Comparative Study of Dynamics Formulation Complexity," *IEEE Transactions on Systems, Man, and Cybernetics,* Vol. 10, November 1980, pp. 730–736.
55. W.A. Khan, V.N. Krovi, S.K. Saha and J. Angeles, "Recursive Kinematics and Inverse Dynamics for a Planar 3R Parallel Manipulator," *ASME Journal of Dynamics Systems, Measurement, and Control,* Vol. 127, No. 4, 2005, pp. 529–536.
56. W.W. Armstrong, "Recursive Solution to the Equations of Motion in an N-Link Manipulator," Proceedings of the 5th World Congress on Theory of Machines and Mechanisms, Vol. 2, Montreal, July 1979.
57. D. Tesar and M. Thomas, "Assessment for the Mathematical Formulation for the Design and Digital Control of Programmable Manipulator Systems," NSF Grant ENG 78-20112.
58. L.J. Chang, "A New Lagrangian Formulation of Dynamics for Robot Manipulators," *Journal of Dynamics Systems, Measurement, and Control,* Volume 111, 1989, pp. 559–567.
59. J.J.E. Slotine and W. Li, "On the Adaptive Control of Robotic Manipulators," *International Journal of Robotics Research,* Vol. 6, No. 3, 1987, pp. 49–59.
60. M.W. Spong, "On the Robust Control of Robot Manipulators," *IEEE Transactions on Automatic Control,* Vol. 37, No. 11, 1992, pp. 1782–1786.
61. A.B. Sharkawy, M.M. Othman and A.M.A. Khalil, "A Robust Fuzzy Tracking Control Scheme for Robotic Manipulators with Experimental Verification," *Intelligent Control and Automation,* Vol. 2, No. 2, 2011, pp. 100–111.
62. M. Galicki, "An Adaptive Regulator of Robotic Manipulators in the Task Space," *IEEE Transactions on Automatic Control,* Vol. 53, No. 4, 2008, pp. 1058–1062.
63. J.H. Hollerbach, Computer Brains and Control Movement, Manchester Institute of Technology, Artificial Intelligence Laboratory, June 1982.
64. A.D. Luca, A. Isidori and F. Nicolo, "Control of Robot Arm with Elastic Joints via Nonlinear Dynamic Feedback," 24th IEEE Conference on Decision and Control, Fort Lauderdale, FL, 1985.
65. J.S Kumar and E.K. Amutha, "Control and Tracking of Robotic Manipulator Using PID Controller and Hardware in Loop Simulation", 2014 International Conference on Communication and Network Technologies (ICCNT), IEEE, 2014, pp. 1-3, DOI: 10.1109/CNT.2014.7062712.
66. Carl Knopse, "PID Control," *IEEE Control System Magazine,* February 2006.
67. J.J. Craig, Introduction to Robotics: Mechanics and Control, Addison-Wesley, Singapore.
68. B.C. Kuo, Automatic Control Systems, 6th Ed., Prentice Hall, Englewood Cliffs, NJ, 1990.
69. K. Ogata, Modern Control Engineering, Prentice Hall of India, New Delhi, 2010.

70. Manjeet and P. Khatri, "Trajectory Control of Two Link Robotic Manipulator Using PID," *Golden Research Thoughts,* Vol. 3, No. 5, November 2013, pp. 1–8.
71. Zuoshi Song, Jianqiang Yi, Dongbin Zhao and Xinchun Li, "A Computed Torque Controller for Uncertain Robotic Manipulator Systems: Fuzzy Approach," *Fuzzy Sets and Systems,* Vol. 154, No. 2, September 2005, pp. 208–226.
72. A.J. Shah and S.S. Ratan, "Dynamics Analysis of Two Link Robot Manipulator for Control Design Using PID Computed Torque Control", *International Journal of Robotics and Automation (IJRA),* Vol. 5, No. 4, December 2016, pp. 277–283.
73. M. A. Llama, R. Kelly and V. Santibañez, "Stable Computed-Torque Control of Robot Manipulators via Fuzzy Self-Tuning," *IEEE Transactions on Systems, Man, And Cybernetics—Part B: Cybernetics,* Vol. 30, No. 1, February 2000, pp. 143–150.
74. K.J. Astrom and T.J. McAvoy, "Intelligent Control," *Journal of Process Control,* Vol. 2, No. 3, 1992, pp. 115–127.
75. G.N. Saridis, "Intelligent Robotic Control," *IEEE Transactions on Automatic Control,* Vol. AC-28, No. 5, 1983, pp. 547–557.
76. K.S. Narendra and S. Mukhopadhyay, "Intelligent Control Using Neural Network," *IEEE Control Systems Magazine,* Vol. 12, No. 2, April 1992, pp. 11–18.
77. K. Lochan, B.K. Roy and B. Subudhi, "A Review on Two-Link Flexible Manipulators," *Annual Reviews in Control,* Vol. 42, 2016, pp. 346–367.
78. L. Meirovitch, Methods of Analytical Dynamics, McGraw-Hill, New York, NY, 1970.
79. L. Meirovitch, Elements of Vibration Analysis, McGraw-Hill, New York, NY, 1970.
80. L. Megdalena and F. Monasterio, "A Fuzzy Logic Controller with Learning through the Evolution of Its Knowledge Base," *International Journal of Approximate Reasoning,* Vol. 16, No. 3-4, 1997, pp. 335–358.
81. C.J. Harris, C.G. Moore and M. Brown, Intelligent Control: Aspects of Fuzzy Logic and Neural Nets, Vol. 6: World Scientific Series in Robotics and Intelligent Systems, World Scientific, Singapore, 1993, pp. 1–33.
82. J.Y. Lai and J.J. Shieh, "On the Development of a Fuzzy Model Based Controller for Robotic Manipulator," Proceedings of the IEEE/RSJ International Conference on Intelligent Robots and Systems, Raleigh, NC, Vol. 1, 1992, 287–292.
83. D.K. Pratihar, Soft Computing Fundamentals and Applications, Narosa Publishing House Pvt. Ltd., India, 2015.
84. G.J. Klier and B.O. Yuan, Fuzzy Sets and Fuzzy Logic Theory and Applications, Prentice Hall, Englewood Cliffs, NJ, 1995.
85. L.X. Wang, "Analysis and Design of Hierarchical Fuzzy System," *IEEE Transactions on Fuzzy Systems,* Vol. 7, No. 5, 1999, pp. 617–624.
86. M.L. Lee, H.Y. Chung and F.M. Yu, "Modelling of Hierarchical Fuzzy Systems, *Fuzzy Sets and Systems,"* Vol. 138, 2003, pp. 343–361.
87. W. Li, "Design of a Hybrid Fuzzy Logic Proportional Plus Conventional Integral-Derivative Controller," *IEEE Transactions on Fuzzy Systems,* Vol. 6, No. 4, 1998, pp. 449–463.
88. W. Li, X.G. Chang, F.M. Wahl and Jay Farrell, "Tracking Control of a Manipulator under Uncertainty by FUZZY P + ID Controller," *Fuzzy Sets and Systems,* Vol. 122, 2001, pp. 125–137.
89. S.G. Anavatti, S.A. Salman and J.Y. Choi, "Fuzzy +PID Controller for Robot Manipulator," International Conference on Computational Intelligence for Modelling Control and Automation and International Conference on Intelligent Agents Web Technologies and International Commerce (CIMCA'06), Sydney, NSW, 2006, p. 75.
90. R. Kelly, R. Haber, R.E. Haber-Guerra, F. Reyes, "Lyapunov stable control of robot manipulators a fuzzy self-tuning procedure", *Intelligent Automation and Soft Computing,* Vol. 5, No. 4, 1999, pp. 313–326.

91. F. Piltan, N. Sulaiman, Arash Zargari, Mohammad Keshavarz and Ali Badri, "Design PID-Like Fuzzy Controller with Minimum Rule Base and Mathematical Proposed On-line Tunable Gain: Applied to Robot Manipulator," *International Journal of Artificial Intelligence and Expert Systems (IJAE),* Vol. 2, No. 4, 2011, pp. 184–194.
92. F. Piltan, et al., "Design Sliding Mode Controller for Robot Manipulator with Artificial Tunable Gain," *Canadian Journal of Pure and Applied Science,* Vol. 5, No. 2, 2011, pp. 1573–1579.
93. F. Piltan, N. Sulaiman, Zahra Tajpaykar, Payman Ferdosali and Mehdi Rashidi, "Design Artificial Nonlinear Robust Controller Based on CTLC and FSMC with Tunable Gain," *International Journal of Robotics and Automation,* Vol. 2, No. 3, 2011, pp. 195–210.
94. F. Piltan, Nasri Sulaiman, A. Gavahian, Samira Soltani and S. Roosta, "Design Mathematical Tunable Gain PID-Like Sliding Mode Fuzzy Controller with Minimum Rule Base," *International Journal of Robotics and Automation,* Vol. 2, No. 3, pp. 146.
95. J. Han, "From PID to Active Disturbance Rejection Control," *IEEE Transactions on Industrial Electronics,* Vol. 56, No. 3, 2009, pp. 900–906.
96. I. Podlubny, "Fractional-Order Systems and $PI^{\lambda}D^{\mu}$ Controller," *IEEE Transactions on Automatic Control,* Vol. 44, No. 1, 1999, pp. 208–214.
97. D. Xue and Y. Chen, "A Comparative Introduction of Four Fractional Order Controllers," Proceedings of the 4th World Congress on Intelligent Control and Automation, Shanghai, China, 2002, pp. 3228–3235.
98. Chun-Tang Chao, Nana Sutarna, Juing-Shian Chiou and Chi-Jo Wang "An Optimal Fuzzy PID Controller Design Based on Conventional PID Control and Nonlinear Factors," *Journal of Applied Sciences,* Vol. 9, 2019, p. 1224.
99. B.G. Hu, G.K.I. Mann and R.G. Gosine, "New Methodology for Analytical and Optimal Design of Fuzzy PID Controllers," *IEEE Transactions on Fuzzy Systems,* Vol. 7, 1999, pp. 521–539.
100. K. Yan, "Application of Fuzzy Control under Time Varying Universe in Unmanned Vehicles," Proceedings of the 33rd Youth Academic Annual Conference of Chinese Association of Automation (YAC), Nanjing, China, 2018, pp. 439–444.
101. J.G. Ziegler and N.B. Nichols, *"Optimum Settings for Automatic Controllers," Transactions of the ASME,* Vol. 64, 1942, pp. 759–765.
102. M.A. Llama, K. Rafael and V. Santibanez, "A Stable Motion Control System for Manipulators via Fuzzy Self-Tuning," *Fuzzy Sets and Systems,* Vol. 124, 2001, 133–154.
103. B.A.M. Wakileh and K.F. Gill, "Use of Fuzzy Logic in Robotics," *Computers in Industry,* Vol. 10, 1988, pp. 35–46.
104. C.M. Lim and T. Hiyama, "Application of Fuzzy Logic Control to a Manipulator," *IEEE Transactions on Robotics and Automation,* Vol. 1, No. 5, 1991, pp. 688–691.
105. S. Agatonovic-Kustrin and R. Beresford "Basic Concepts of Artificial Neural Network (ANN) Modelling and Its Application in Pharmaceutical Research," *Journal of Pharmaceutical and Biomedical Analysis,* Vol. 22, No. 5, 2000, pp. 717–727.
106. J.A. Anderson and Edward Rosenfeld, Neurocomputing: Foundations of Research, MIT Press, Cambridge, MA, 1988.
107. B. Muller, J. Reinhardt and M.T. Strickland, Neural Networks: An Introduction, Springer-Verlag, Berlin, 1995.
108. P. Gupta and N.K. Sinha, "Control of Robotic Manipulators Using Neural Networks – A Survey," in S.G. Tzafestas (ed.), Methods and Applications of Intelligent Control, Kluwer Academic Publishers, Amsterdam, 1997, pp. 103–136.
109. F.L. Lewis, A. Yesildirek and K. Liu, "Multilayer Neural-Net Robot Controller with Guaranteed Tracking Performance," *IEEE Transactions on Neural Networks,* Vol. 7, 1996, p. 388.

110. Y.H. Kim and F.L. Lewis, "Optimal Design of CMAC Neural-Network Controller for Robot Manipulators, IEEE Transactions on Systems, Man, and Cybernetics, Vol. 30, 2000, p. 22.
111. A.J. Koivo, Fundamentals for Control of Robotic Manipulators, Wiley, New York, NY, 1989.
112. T.Y. Kuc and W.G. Han, "Adaptive PID Learning of Periodic Robot Motion," IEEE Conference on Decision and Control, 1998, p. 186.
113. F.L. Lewis, C.T. Abdallah and D.M. Dawson, Control of Robot Manipulators, Macmillan, New York, NY, 1993.
114. S.M. Prabhu and D.P. Garg, "Artificial Neural Network Based Robot Control: An Overview," *Journal of Intelligent and Robotic Systems,* Vol. 15, 1996, pp. 333–365.

# 6 Maize Leaf Disease Detection and Classification Using Deep Learning

*Phani Kumar Singamsetty*
GITAM University Hyderabad, India

*G. V. N. D. Sai Prasad*
GITAM University Hyderabad, India

*N. V. Swamy Naidu*
National Institute of Technology Raipur, India

*R. Suresh Kumar*
National Institute of Technology Raipur, India

**CONTENTS**

## 6.1 INTRODUCTION

*Zea mays* L., which is the binomial name for maize, is the most cultivated cereal crop globally, surpassing the production of rice and wheat, and is recognized as the queen of cereal crops. Maize is cultivated in about 190 million hectares across 165 countries with a 39% share of global grain production. Maize is the third largest crop produced in India after rice and wheat. It is predominantly a Kharif season crop, which accounts for 10% of total food grain production in India. Although the direct consumption of maize by humans is less, it is being used as a raw ingredient in a wide range of products such as oil, starch, protein, alcoholic beverages, cosmetics, textile, food sweeteners, gum, packaging and paper industries. Hence, there is a dire need for smart farming method for early detection and control of crop diseases in maize to produce better yield. Deep learning-based models such as the convolutional neural network (CNN) have been found to pattern better the type of disease from the input images with a good classification accuracy [1-12]. Some of the commonly found corn diseases such as grey leaf spot (GLS), corn rust and northern leaf blight, which have a profound impact on the yield per hectare, are presented in this chapter. The causes of diseases, early symptom detection and their impact on the yield per hectare have been thoroughly discussed. The diseased leaf is compared with healthy maize leaf shown in Figure 6.1 for the present study.

## 6.2 TYPES OF MAIZE LEAF DISEASES

### 6.2.1 Grey Leaf Spot

GLS, known usually as Cercospora leaf spot, is a corn disease in which maize leaf is found to have grey spots. It is a common fungal disease. The picture of GLS diseased maize leaf is shown in Figure 2(a). GLS is a common disease in the maize crop which is caused by a pathogen *Cercospora zeae-maydis*, hence the name Cercospora leaf spot. Warm temperatures and high humidity conditions are more favourable for the spreading of this GLS disease [1, 2]. The spreading rate of this disease is more towards the end of the summer season. Grey spots begin to appear as small necrotic spots with chlorotic halos, with colourations from tan to brown beginning to appear before sporulation. As the infection progresses, lesions begin to grow and are restricted by parallel veins in the leaves. As GLS progresses, lesions will coalesce and form larger necrotic areas. GLS lesions hinder the photosynthesis process,

**FIGURE 6.1** Healthy maize leaf.

thereby reducing the carbohydrate supply to the grain filling and adversely impacting the crop yield. Studies show that crop yield may be reduced by 2 to 50% due to spread of GLS disease in the maize leaves.

### 6.2.2 Common Rust

Common rust [2] is a fungal pathogen disease in corn caused by *Puccinia sorghi*. The spread of this disease is favoured by moist and cool conditions. Hot and dry conditions stop or slow down the spread of this disease. Maize leaf infected with common rust is shown in Figure 6.2(b). Common rust spreads through windblown spores. Corn rust disease lesions reduce the functional area of the leaves for photosynthesis, thereby reducing sugars. This reduction in sugars causes the kernels to fill up the required sugars from the stalk. Due to this, stalks will be weakened and this in turn increases the potency for the stalk to rot. Significant yield loss occurs with poorly filled kernels. The disease is noticed with small tan spots appearing on the corn leaves which are in brick red colour and are visible in jagged appearance.

### 6.2.3 Northern Leaf Blight

Northern leaf blight [3-8] is a foliar type of crop disease in corn caused by a fungus *Exserohilum turcicum*, which is actively found in humid weather conditions. Maize leaf infected with northern leaf blight disease is shown in Figure 6.2(c). It survives in the corn debris and builds over time, which even has a tendency to infect the new season's crop thereafter. High dew, regular rains, higher humidity and moderate temperate conditions favour the spread of this disease. This disease will spread to new leaves through spores blown through rain splash and air currents. Early symptoms are visible in 1 or 2 weeks after the infection, which are in the form of grey-green elliptical lesions. As time elapses, these lesions become pale grey to tan and enlarge to sizes ranging from 1 to 6 inches or larger. We can easily identify northern leaf blight lesions, which are cigar-shaped and not restricted by the veins. These lesions grow and coalesce and spread across the entire leaf and leave a grey or burned appearance. Decreased photosynthesis will result in yield loss. Northern leaf blight usually starts from lower leaves of the plant and spreads to the upper portion leaves. If the upper leaves are affected by this disease during pollination or early ear fill, there will be significant yield loss.

## 6.3 LITERATURE REVIEW

Zhang et al. employed genetic algorithm (GA) coupled with support vector machines (SVMs) for classifying various diseases in maize leaf [4]. Brahimi et al. also carried out a study on classifying tomato diseases [7]. To classify maize leaf disease and

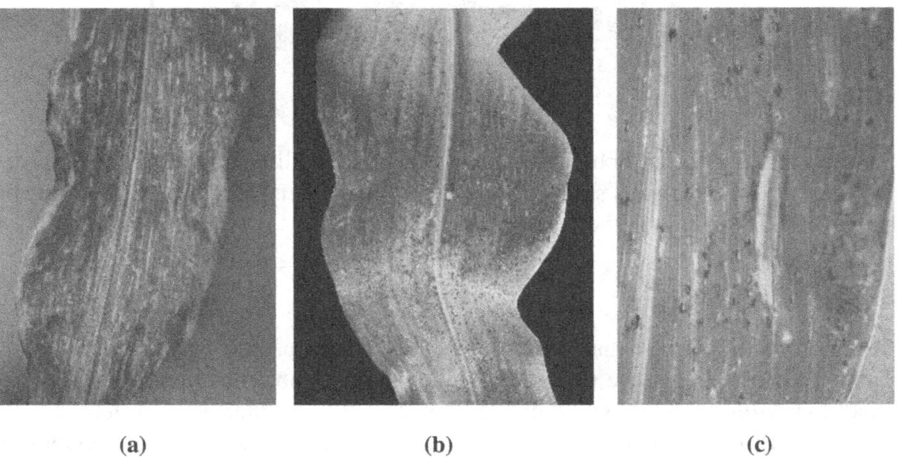

(a) (b) (c)

**FIGURE 6.2** Infected maize crop leaf. (a) Grey leaf spot (GLS). (b) Common rust. (c) Northern leaf blight disease.

compare diseased leaf with healthy leaf, Alehegn succeeded in developing a technique which is dependent on colour, texture and structural feature [8]. Ahmadi et al. used the artificial neural network (ANN) for classifying infections in oil palm trees by fungus [9]. Sibiya and Sumbwanyambe successfully used CNN for maize leaf disease classification and claimed to have attained an accuracy of 92.85% using this technique [10]. Priyadharshini et al. used LeNet architecture for classifying maize leaf diseases and found the accuracy in using the architecture to be 97.89% with training over 1,000 epochs [11]. The research showed several working solutions for maize leaf disease classification using 1,000 epochs for training which needs high computational power and time [12-15]. To overcome these drawbacks and obtain higher efficiencies with less computational time, a pretrained neural network, AlexNet, with transfer learning has been found suitable. The present work is aimed at using CNN and the transfer learning-based pre-trained network such as AlexNet for identification of crop diseases [12]. The work presented in the chapter is organized into two sections. In the first section, we present materials and methods and development of experimental setup for presenting the basic results. Finally, the last section deals with critical evaluation of the proposed models in comparison with the existing techniques.

## 6.4 MATERIALS AND METHODS

### 6.4.1 Materials

Maize crop leaf disease data in the form of images have been collected from Plant Village Dataset [15]. Images distribution in various classes of the training and the test dataset are shown in Table 6.1. The classes are more or less balanced in terms of images in each class.

### 6.4.2 Methods

A custom CNN architecture is designed by training it on the trained data set. CNN architecture procedure and layers involved are presented in Figure 6.3 for intuition.

#### 6.4.2.1 Custom CNN Layers

The custom CNN starts with an image input with a three-channel image (i.e. image in RGB with a size of 256 × 256 pixels). The present CNN is made with seven-layer

**TABLE 6.1**
**Distribution of Maize Leaf Images in Training Data Set and Test Data Set**

| | No. of Images | |
|---|---|---|
| **Class** | **Training Data Set** | **Test Data Set** |
| Cercospora leaf spot or GLS | 1,642 | 410 |
| Common rust | 1,907 | 477 |
| Northern leaf blight | 1,908 | 465 |
| Healthy | 1,859 | 477 |

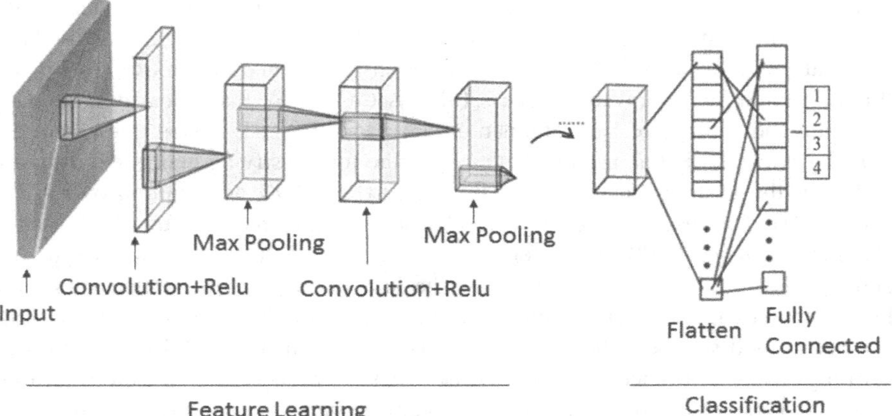


**FIGURE 6.3** Architecture of convolutional neural network (CNN).

convolutions with every single layer followed by a batch or section normalization layer and a max pooling layer, while the last convolutional layer has no max pooling layer but is fully connected. Each convolutional layer has neurons numbered 8, 16, 32, 64, 128, 256, and 512, respectively, and every convolutional layer has same stride [1 1], with size of each convolutional layer being 3 × 3 × number of neurons from the previous layer. For instance, the second convolutional layer will be of size 3 × 3 × 8, since the first convolutional layer has 8 neurons. All the batch or section normalization layers are followed by convolutional layer comprising same number of channels as that of neurons, i.e. for the first layer, the batch normalization layer is 8-channel layer equalling the neurons in the corresponding layer. Every batch normalization layer is followed by a converging max pooling layer having same stride [2 2]. The last layer having no max pooling layer is fully connected and has four neurons. The output of the classification between the four classes and the last connected layer has softmax activation, while all the remaining convolutional layers have an activation ReLU (rectified linear unit).

#### *6.4.2.1.1 Convolutional Layers (CL)*

Kernel convolution has been used in CNN where it becomes a key element for many other algorithms for computer vision. It is a method in which a small number matrix (called the kernel or filter) is taken and convolved over the image and is transformed based on the filter values. The feature map values are determined using equation (1).

$$G[m,n]=(f*h)[m,n]=\sum_{j}\sum_{k}h[j,k]\; f[m-j,n-k] \tag{1}$$

where the input images are indicated by $f$ and kernel is denoted by $h$. The indices of rows and columns of the resultant matrix are marked with $m$ and $n$, respectively. Summations are done over number of columns and number of rows of input image.

#### 6.4.2.1.2 *Pooling Layer*

Pooling of layer reduces the dimensions of the data by taking a single value (sum/average/maximum) from a feature map covered by the kernel.

**Sum pooling:** If the sum of all the values of a feature map covered by the kernel is taken, it is known as sum pooling

**Average pooling:** If the average value of all the values of a feature map covered by the kernel is taken, it is called average pooling.

**Max Pooling:** If maximum value of all the values of a feature map covered by the kernel is taken, it is called max pooling. Equation (2) shows the procedure for the calculation for max pooling operation. In the present proposed custom CNN design, the max pooling technique has been used.

$$h[i,j] = \max\{X_{i+k,\ j+l-1,}\ 1 \le k \le m \text{ and } 1 \le l \le m\} \tag{2}$$

#### 6.4.2.1.3 *Batch Normalization Layer*

This is a technique for deep NN training that standardizes the inputs to a layer for each mini-batch, which in turn affects the stability of the learning process by substantially reducing the amount of training epochs needed to train deep networks.

$$\frac{\bar{\mathrm{X}} = \mathrm{xi} - \mu\mathrm{B}}{\sqrt{\sigma^2 + €}} \tag{3}$$

where xi is the input image, μB is the batch mean and $\sigma^2$is the batch variance.

#### 6.4.2.1.4 *ReLU*

The rectified linear activation is a function based on a very meek calculation which returns the provided value as input directly or the value 0.0 if the input 0.0 or less (Eq. (4)).

$$g(z) = \max(0, x). \ldots \tag{4}$$

### 6.4.2.2 Working Principle

The first CL extracts miniature features such as edges, lines, curves, colours and blobs, while the next one learns the global features. Back-propagation is the learning technique behind the CNN training process whose algorithm comprise four well-defined parts, namely the forward pass, the loss function, the backward pass and the weight update. In the forward pass, the CNN takes the training image, which passes through the complete network. All the filter values are randomly initialized, which may not be able to learn low-level features. Thus, based on the loss function taken during the training stage, weights will be reinitialized during the backward pass. The above four well-defined parts are employed for every iteration. This is continued for a finite number of iterations or until the attained threshold limit of the loss function. Training may be stopped if there is no improvement in the validation accuracy for certain pre-set number of iterations.

#### 6.4.2.3 Transfer Learning

Transfer learning is a methodology in which the knowledge gained by a CNN for classification of a particular dataset can be used later on some other new dataset type. The knowledge is nothing but the weights updated during training. Initial layers of CNN basically learn about higher level features such as edges, curves and colours. This learning will also be useful for any other dataset. In transfer learning, initial layers' weights of pre-trained CNN can be frozen and only the lower level layers are allowed to update their weights based on specific low-level features of new dataset.

#### 6.4.2.4 AlexNet Architecture

AlexNet architecture shown in Figure 6.4 forms the basis. AlexNet [14] consists of five CLs coupled with three full connected layers, with multiple CLs extracting the salient features of an image. AlexNet is made in such a way that it contains different kernels in the same CL. The first two CLs are followed by a converging max pooling layer, and the next layers are the CLs directly connected whose output is connected to an overlapping max pooling layer. The output of this layer is connected to a series of three fully connected layers and the last layer with softmax activation having 1,000 neurons, which implies that AlexNet can successfully

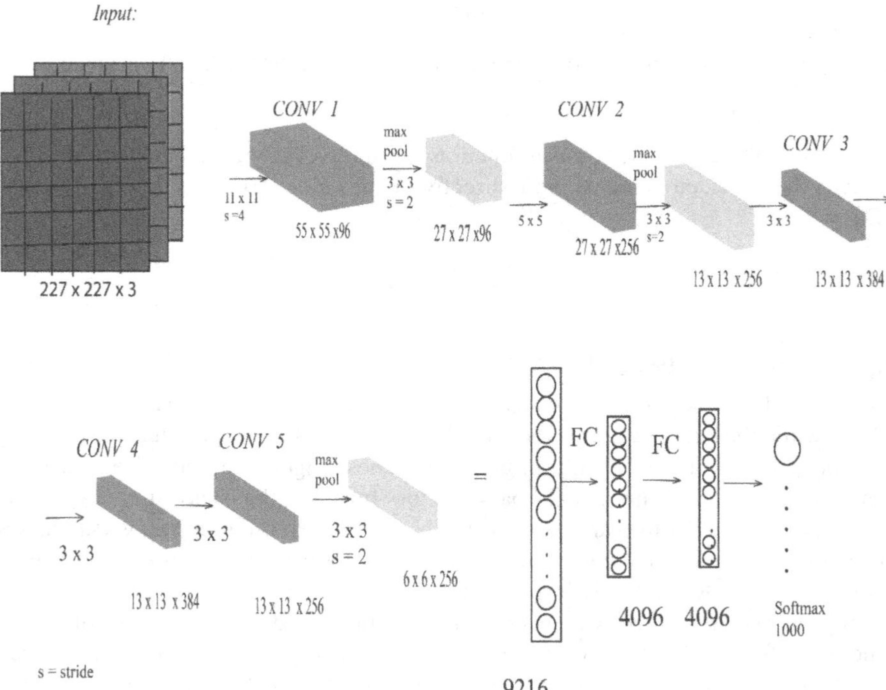


**FIGURE 6.4** Architecture of AlexNet.

classify 1,000 objects. After convolution and fully connecting the layers, nonlinearity of ReLU is added. Before performing pooling, the nonlinearity of the first and second convolution layers in the ReLU is followed by a local normalization step. Max pooling layers are usually used to downsample the tensor width and height while maintaining the same depth. Max pool layers overlapping either with similar max pool layers or with the neighbouring frames over which the max is measured overlap each other. The authors used size 3 range window pooling between the adjacent windows with a move of 2. This concurrent aspect of pooling helped in reducing the error by 4.6%. The use of ReLU non-linearity is an important feature of AlexNet. Activation function tangent hyperbolic (tanh) or sigmoid is being used as the standard way to train the model of the neural network. However, AlexNet showed that deep CNNs could be trained much faster by the use of nonlinearity of ReLU than the standard saturating activation functions such as tanh or sigmoid. In transfer learning technique using AlexNet, initial 10 layers of AlexNet are frozen and last 2 fully connected layers are replaced by a fully connected layer with softmax activation with four neurons representing four output classes and trained it on the training data set.

## 6.5 EXPERIMENTATION AND RESULTS

The proposed CNN model has been employed for maize crop leaf disease classification and detection.

### 6.5.1 Experimental Setup

All experiments are done using the following software and hardware setup.

| | | |
|---|---|---|
| Software | : | Matlab2019b |
| Operating System | : | Windows 10 |
| Hardware | : | Nvidia GeForce RTX 2070 Super 8GB GPU with 32 GB of RAM |

### 6.5.2 Training

CNN is trained on the training dataset with 10 epochs and 114 iterations per epoch. Minimum batch size is set at 64 due to memory constraints. Ten-fold cross-validation has been considered while training CNN. Validation frequency has been set to 114 iterations with infinite patience. The start learning rate is set at 0.002 with Stochastic Gradient Descent with Momentum (SGDM) optimizer of momentum 0.9, and L2 regularization of $1.0e^{-04}$ has been used. A validation accuracy of 98.50% is obtained at the end of the training. The training progress on custom-designed CNN with validation accuracy is shown in Figure 6.5. Pre-trained AlexNet with last layers modified train on training dataset. While training AlexNet transfer learning randomly, 80% of training images have been used for training and remaining 20% are used for validation. Validation accuracy of 100% is achieved by the end of training. Training options were set as following:

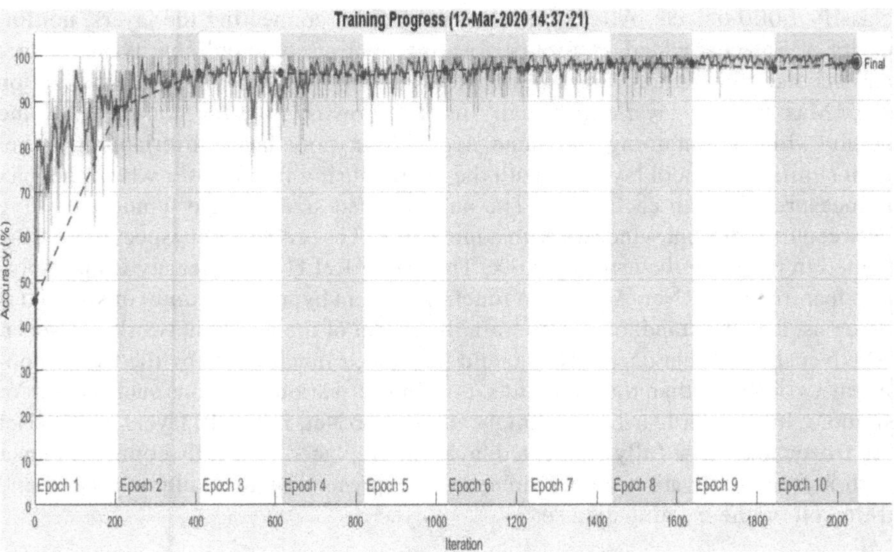


**FIGURE 6.5** Training progress on custom CNN.

minimum batch size is 32, validation frequency is set to 228 iterations with SGDM optimizer with a momentum of 0.9 and L2 regularization of $1.0e^{-04}$ is used. While training the AlexNet, initial 10 layers are frozen, which indicates that the weights for these layers are constant and do not change with training. Training progress with validation accuracy is shown in Figure 6.6.

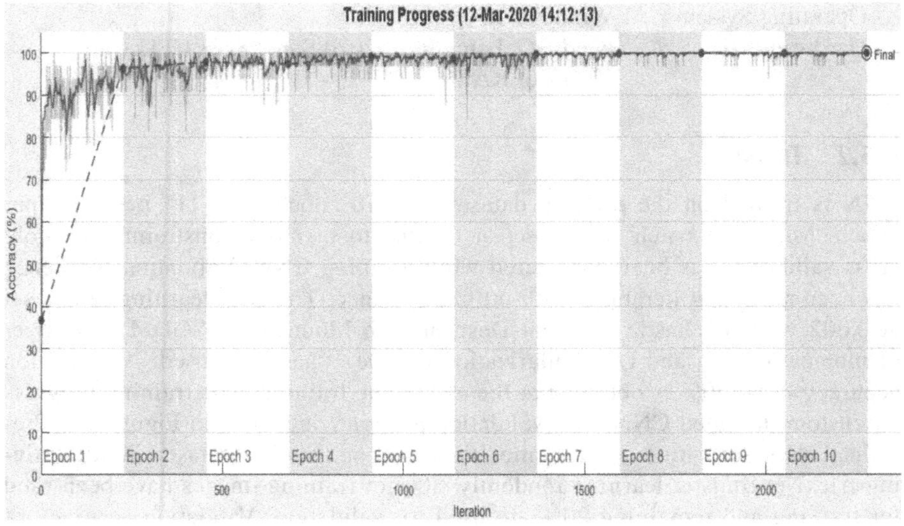


**FIGURE 6.6** Training progress with AlexNet transfer learning.

### 6.5.3 Testing

Testing is done on a separate test dataset and the images in the test set are totally unseen by CNN while training. Trained AlexNet is able to better generalize over the test images, and classification accuracy of **97.81%** has been achieved while testing.

## 6.6 CRITICAL ANALYSIS OF RESULTS AND DISCUSSION

### 6.6.1 Custom CNN Results

Table 6.2 enlists the confusion matrix after classification on test data using custom-designed CNN, while Table 6.3 shows hits, misses, correct rejections and false alarms which help in finding the evaluation metrics such as accuracy, sensitivity, specificity, precision, recall and F1 score. Table 6.4 shows class-wise accuracy, sensitivity, specificity, precision, recall and F1 score calculated from data shown in Table 6.3 for the proposed CNN.

**True positive (TP) or hit:** An outcome-based model correctly predicting the classes.

**True negative (TN) or correct rejection:** This outcome model correctly predicts the negative class.

**False positive (FP) or false alarm or type-I error:** The outcome-based model that incorrectly predicts the positive class.

**False negative (FN) or miss or type-II error:** The outcome-based model that incorrectly predicts the negative class.

**Accuracy:** It is the ratio of the correctly labelled subjects to the whole pool of subjects.

$$\text{Accuracy} = \frac{\text{TP} + \text{TN}}{\text{TP} + \text{FP} + \text{TN} + \text{FN}} \tag{5}$$

Overall classification accuracy: **97.59%**

**Precision:** It is the ratio of the correctly positively labelled objects by our program to all positively labelled objects.

$$\text{Precision} = \frac{\text{TP}}{\text{TP} + \text{FT}} \tag{6}$$

**TABLE 6.2**
**Confusion Matrix after Classification Using Proposed Custom-Designed CNN**

| Actual/Predicted | Grey Leaf Spot | Common Rust | Northern Leaf Blight | Healthy |
|---|---|---|---|---|
| Grey leaf spot | 384 | 0 | 26 | 0 |
| Common rust | 1 | 476 | 0 | 0 |
| Northern leaf blight | 15 | 1 | 460 | 1 |
| Healthy | 0 | 0 | 0 | 465 |

**TABLE 6.3**
**Hits, Correct Rejections and Errors like False Alarm and Miss**

| Class Name | TP (Hit) | FN (Miss) | TN (Correct Rejection) | FP (False Alarm) |
|---|---|---|---|---|
| Grey leaf spot | 384 | 26 | 1,403 | 16 |
| Common rust | 476 | 1 | 1,351 | 1 |
| Northern leaf blight | 460 | 17 | 1,326 | 26 |
| Healthy | 465 | 0 | 1,363 | 1 |

**Sensitivity:** It is the ratio of correctly classified objects which are really correct

$$\text{Sensitivity} = \frac{\text{TP}}{\text{TP} + \text{FN}} \tag{7}$$

**Specificity:** It is the ratio of correctly classified negative objects as negative.

$$\text{Specificity} = \frac{\text{TN}}{\text{TN} + \text{FN}} \tag{8}$$

**F1 score:** It is the harmonic mean of precision and sensitivity.

$$\text{F1 score} = \frac{2 * \text{TP}}{(2 * \text{TP} + \text{FP} + \text{FN})} \tag{9}$$

Different metrics for classification using custom CNN are shown in Table 6.6. The accuracies for each class are better than the existing methods developed by Ahmadi et al. and Sibiya and Sumbwanyambe [9, 10]. The maximum overall error percentage in our proposed CNN model is only 2.41%. Class-wise accuracy is also very astonishing: the maximum error percentage in each class is only 2%.

### 6.6.2 Results Using the AlexNet

A different model with the transfer learning technique using a pre-trained neural network such as AlexNet has also been proposed and presented in the present study.

**TABLE 6.4**
**Class-Wise Accuracy, Sensitivity, Specificity, Precision, Recall and F1 Score**

| Class Name | Accuracy | Sensitivity | Specificity | Precision | Recall | F1score |
|---|---|---|---|---|---|---|
| Grey leaf spot | 97.70 | 93.66 | 98.87 | 96.00 | 93.66 | 94.81 |
| Common rust | 99.89 | 99.79 | 99.93 | 99.79 | 99.79 | 99.79 |
| Northern leaf blight | 97.65 | 96.44 | 98.08 | 94.65 | 96.44 | 95.53 |
| Healthy | 99.95 | 100.00 | 99.93 | 99.79 | 100.00 | 99.89 |

All the evaluation metrics generated with this network are presented below in detail. Table 6.5 presents the confusion matrix generated from classification on test dataset using AlexNet and transfer learning. Hits, correct rejections and errors like false alarm and miss values calculated from confusion matrix in Table 6.5 are shown in Table 6.6. Table 6.6 depicts the true positives and false negatives, which helps in finding the evaluation metrics such as accuracy, sensitivity, specificity and F1 score. Class-wise accuracy, sensitivity, specificity, precision, recall and F1 score calculated from data in Table 6.6 are shown in Table 6.7 for the proposed transfer learning on AlexNet. As shown in Table 6.7, it can be seen that there is some improvement in class-wise accuracies, and an overall accuracy in other metrics, using AlexNet with transfer learning.

**TABLE 6.5**
**Confusion Matrix after Classification Using Transfer Learning on AlexNet**

| Actual/Predicted | Grey Leaf Spot | Common Rust | Northern Leaf Blight | Healthy |
|---|---|---|---|---|
| Grey leaf spot | 389 | 1 | 20 | 0 |
| Common rust | 0 | 476 | 1 | 0 |
| Northern leaf blight | 15 | 3 | 459 | 0 |
| Healthy | 0 | 0 | 0 | 465 |

**TABLE 6.6**
**Hits, Correct Rejections and Errors like False Alarm and Miss**

| Class Name | TP (Hit) | FN (Miss) | TN (Correct Rejection) | FP (False Alarm) |
|---|---|---|---|---|
| Gray leaf spot | 389 | 21 | 1,404 | 15 |
| Common rust | 476 | 1 | 1,348 | 4 |
| Northern leaf blight | 459 | 18 | 1,331 | 21 |
| Healthy | 465 | 0 | 1,364 | 0 |

**TABLE 6.7**
**Class-Wise Accuracy, Sensitivity, Specificity, Precision, Recall and F1 Score**

| Class Name | Accuracy | Sensitivity | Specificity | Precision | Recall | F1 Score |
|---|---|---|---|---|---|---|
| Gray leaf spot | 98.03 | 94.88 | 98.94 | 96.29 | 94.88 | 95.58 |
| Common rust | 99.73 | 99.79 | 99.70 | 99.17 | 99.79 | 99.48 |
| Northern leaf blight | 97.87 | 96.23 | 98.45 | 95.63 | 96.23 | 95.92 |
| Healthy | 100.00 | 100.00 | 100.00 | 100.00 | 100.00 | 100.00 |

*Overall classification accuracy*: **97.81%**.

**TABLE 6.8**
**Comparison of Proposed Deep Learning Models**

| Class Name | Existing Method 1 (Ahmadi et al.) [9] | Existing Method 2 (Priyadharshini et al.) [11] | Proposed CNN | Transfer Learning on AlexNet |
|---|---|---|---|---|
| Grey leaf spot | 91.00 | 84.58 | 97.70 | 98.03 |
| Common rust | 87.00 | 99.87 | 99.89 | 99.73 |
| Northern leaf blight | 99.90 | 98.14 | 97.65 | 97.87 |
| Healthy | 93.50 | 99.70 | 99.95 | 100.00 |
| Overall | 92.85 | 95.57 | 97.59 | 97.81 |

## 6.7 RESULTS COMPARISION WITH EXISTING METHODS

### 6.7.1 Class-Wise Accuracies

Table 6.8 shows the performance comparison of the proposed deep learning models with the existing methods. The performance comparison of the existing methods in terms of class-wise accuracies is plotted as a bar diagram in Figure 6.7. However, Priyadharshini et al. [11] specified that the accuracy of 97.89% has been achieved using $3 \times 3$ matrix kernel, but evaluation metrics for classification has not been produced for this result. Hence, an evaluation metrics of $5 \times 5$ matrix kernel has been considered in the present study for comparison.

### 6.7.2 Overall Accuracies

Comparison of overall classification accuracies of different methods proposed by different researchers is made with our proposed models. As shown in Table 6.9, the

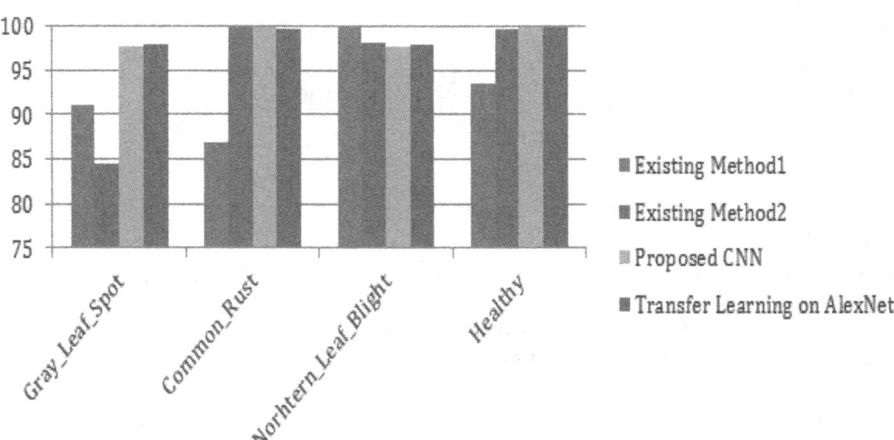


**FIGURE 6.7** Comparison of class-wise accuracies of the proposed models with the existing methods.

**TABLE 6.9**
**Overall Classification Accuracy Comparison of Proposed and Existing Methods**

| Method | Overall Classification Accuracy |
|---|---|
| GA-SVM [4] | 92.82 |
| ANN Classifier [8] | 94.40 |
| SVM [2] | 89.60 |
| Existing CNN [10] | 92.85 |
| LeNet (Kernel5x5) [11] | 95.57 |
| **Proposed CNN** | **97.59** |
| **AlexNet** | **97.81** |

present results pertaining to overall classification accuracies are dominant over the previously proposed models. The same can be visualized graphically and are displayed in Figure 6.8.

## 6.8 CONCLUSION AND FUTURE SCOPE

In this research work, classification of maize leaf diseases is done using custom-designed CNN and transfer learning-based pre-trained neural network AlexNet. Classification accuracy of 97.81% has been achieved using the proposed deep learning models. As a part of smart farming, these trained models can be deployed in

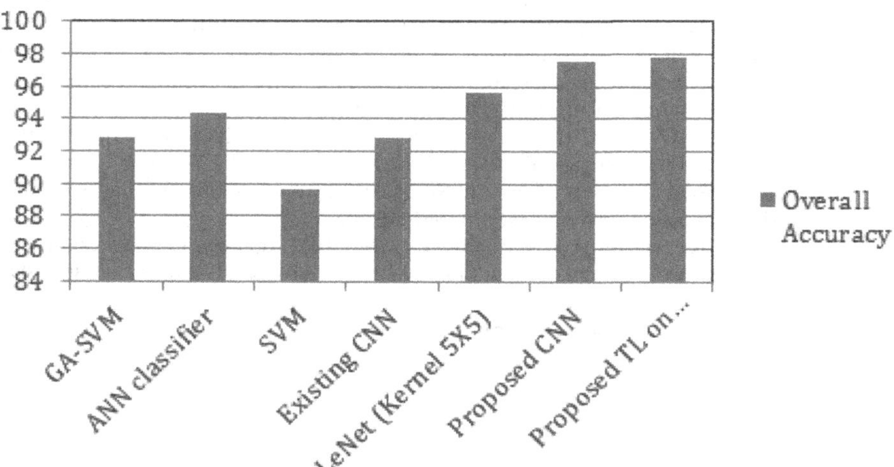


**FIGURE 6.8** Overall classification accuracy comparison of proposed deep learning models with the existing ones.

the future into the edge devices, as this can help farmers detect maize leaf diseases very early and timely action can be suggested by agricultural scientists to control the diseases. Transfer learning technique has also been suitably applied on other latest pre-trained networks to improve upon the classification accuracy.

## REFERENCES


1. Ward, J. M. J., Stromberg, E. L., Nowell, D. C., and Nutter, F. W. Jr. (1999) Grey leaf spot: a disease of global importance in maize production. *Plant Dis* 83:884–895.
2. Zhang, L., and Yang, B. (2014) Research on recognition of maize disease based on mobile internet and support vector machine technique. *Adv Mater Res* 905:659–662. https://doi.org/10.4028/www.scientific.net/AMR.905.659
3. Indian Council of Agricultural Research. Inoculation Methods and Disease Rating Scales for Maize Diseases; 2012 Revised. New Delhi, India: Directorate of Maize Research, 2016.
4. Zhang, Z., He, X., Sun, X., Guo, L., Wang, J., and Wang, F. (2015) Image recognition of maize leaf disease based on GA-SVM. *Chem Eng Trans* 46:199–204.
5. Ren, J. (2012) ANN vs. SVM: which one performs better in the classification of MCCs in mammogram imaging. *Knowl Based Syst* 26:144–153.
6. Jafari, I., Masihi, M., and Zarandi, M. N. (2018) Scaling of counter-current imbibition recovery curves using artificial neural networks. *J Geophys Eng* 15(3):1062–1070.
7. Brahimi, M., Boukhalfa, K., and Moussaoui, A. (2016) Deep learning of tomato diseases: classification and symptoms visualization. *Appl Artif Intell* 31(4):299–315. https://doi.org/10.1080/08839514.2017.1315516
8. Alehegn, E. (2017) Maize leaf diseases recognition and classification based on imaging and machine learning techniques. *Int J Innov Res Comput Commun Eng* 5(12):1–11
9. Ahmadi, P., Muharam, F. M., Ahmad, K., Mansor, S., and Seman, I. A. (2017) Early detection of Ganoderma basal stem rot of oil palms using artificial neural network spectral analysis. *Plant Dis* 101:1009–1016.
10. Sibiya, M., and Sumbwanyambe, M. (2019). A computational procedure for the recognition and classification of maize leaf diseases out of healthy leaves using convolutional neural networks. *Agri Eng* 1:119–131. 10.3390/agriengineering1010009.
11. Priyadharshini, R. A., Arivazhagan, S., Arun, M., and Mirnalini, A. (2019) Maize leaf disease classification using deep convolutional neural networks. *Neural Comput Appl* 31:8887–8895. https://doi.org/10.1007/s00521-019-04228-3
12. Yang, L., Yi, S., Zebg, N., Liu, Y., and Zhang, Y. (2017) Identification of rice diseases using deep convolutional neural networks. *Neurocomputing* 267:378–384
13. NebGuide. Rust disease of corn in Nebraska. 2016. Available online at: http://extensionpublications.unl.edu/assets/pdf/g1680.pdf
14. Krizhevsky, A., Sutskever, I., and Hinton, G. E. (2017). ImageNet classification with deep convolutional neural networks. *Commun ACM* 60(6):84–90. doi:10.1145/3065386. ISSN 0001-0782
15. https://github.com/spMohanty/PlantVillage-Dataset

# Section III

## Application towards Industrial Engineering

# 7 Inspectorate Patterns for Cell Recognition in Cellular Manufacturing

*K. V. Durga Rajesh*
Koneru Lakshmaiah Education Foundation
Vaddeswaram, India

*Venna Gowtham Kumar*
Koneru Lakshmaiah Education Foundation
Vaddeswaram, India

*G. Naga Sai Ram*
Koneru Lakshmaiah Education Foundation
Vaddeswaram, India

*Tanya Buddi*
GRIET
Hyderabad, India

**CONTENTS**

## 7.1 CELLULAR MANUFACTURING

Today's manufacturing field requires producing high-quality and optimum-price products in order to face the competition in the market. Solid marketing strategy is required by the companies in order to meet the challenges and to stay in the competition. Many factors play a key role in improving the productivity, and layout planning is one among them. Optimum and efficient utilization of resources is very important for a company. Therefore, design of layout is a crucial part in increasing productivity and reducing transportation costs. Moreover, in automobile industry, adoption of automated assembly systems is required to produce a large number of products. Thus, in order to implement these systems, grouping of similar parts is important. Therefore, group technology plays a crucial role in this type of grouping application.

In 1920 when product-oriented departments were used to manufacture standardized products in machine companies, which led to reduce transportation; this was considered as the start of group technology. Group technology is one of the best techniques to manufacture a wide variety of products with low demand. Components are categorized in this system based on similar features, which are then created together. Simply, group technology is characterized as identification and grouping of similar parts to take advantage of their similarities in design and manufacturing. Parts' similarities allow them to be classified into part families. Clustering parts and equipment into cells results in cost savings in installation time, lead time and time and process handling. Group technology applies cellular manufacturing. This works on the principle of similar and identical components being generated simultaneously.

Nowadays, customers expect variety of products in short period of time. To meet these requirements, the manufacturing industries need to produce small batches of large variety of products in shorter lead times. This leads to cellular production approach. Cellular production system has attracted considerable interest in the short-term manufacture of innovative products. Countries such as Japan use this technique of cellular production to achieve the technique of just-in-time. Cellular manufacturing contributes to shaping and running manufacturing cells. From cells, machine groups and part family components will be obtained. Cellular production system is understood as a replacement for the traditional system of manufacturing workshops. Cellular manufacturing's goal is to reduce set-up and flow times, further reducing lead times for inventory and production. Several studies have reported detailed reviews of the research involving finding near-optimal solutions using mathematical programming, heuristic, metaheuristic and artificial intelligence. Various optimization methods have been studied to establish a taxonomic structure for a variety of development strategies of various cell formation issues.

Cellular manufacturing operates on the theory of 'similar parts are manufactured similarly and simultaneously'. Hence, by this context, similarity includes similarity in size, shape and manufacturing attributes. The main advantage of the cellular manufacturing is that, instead of production set-ups and planning for single component, the planning can be done entirely for similar products or components. Cellular manufacturing goes beyond the process of grouping the machines and parts.

## 7.2 TERMS ASSOCIATED WITH CELLULAR MANUFACTURING

Different terminologies are used in cellular manufacturing to describe the process associated with manufacturing.

### 7.2.1 Cell

A cell is a place where different machines are grouped together and operations are formed. When machines are arranged in this manner, it is known as cellular layout. Cell is the basic terminology in cellular manufacturing. A cell may contain group of similar or dissimilar machines.

### 7.2.2 Part Family

A group of components that are similar in terms of manufacturing and are to be processed on different machines is known as part family. The parts that are under part family may not be completely similar. They may have two or more operations in common. If two parts have all the common operations, then they are said to be high similarity parts.

### 7.2.3 Intracellular Movement

The movement of the component within the cell is known as intercellular movement. Less intracellular movements are preferred in cellular manufacturing system.

### 7.2.4 Intercellular Movement

The movement of the component between the cells is known as intercellular movement. There should be less intercellular movements in order to reduce machine handling cost.

### 7.2.5 Manufacturing Lead Time

It is the total time taken by the plant or industry to complete the processing of the given data.

### 7.2.6 Throughput Time

Time or period required to process the given part through manufacturing process only is called throughput time.

### 7.2.7 Cycle Time

It is the time between the processing or assembly of the work unit and the beginning of the next task.

## 7.3 BENEFITS OF CELLULAR MANUFACTURING

Several benefits are there in cellular manufacturing than conventional layout. Some of them are outlined and analyzed in detail in this section. The benefits of cellular manufacturing are as follows.

### 7.3.1 Reducing Material Handling Cost

Material handling cost is defined as the cost per unit for moving the part between machines. The transportation cost involves high ratio of costs in the industry in conventional layout. Figure 7.1 illustrates traditional layout where large transportation of parts is observed.

Material handling cost alone constitutes up to 25% in this type of production. Therefore, improvement in conventional layout includes elimination of transportation costs as much as possible and improvement of productivity. Cellular manufacturing plays a vital role in reducing transportation costs. In the case of cellular manufacturing systems, there is minimal transportation of materials, which in turn reduces material handling cost.

Figure 7.2 shows cellular manufacturing layout where there is minimal transportation cost. Even though the flow of material takes place, it is minimal

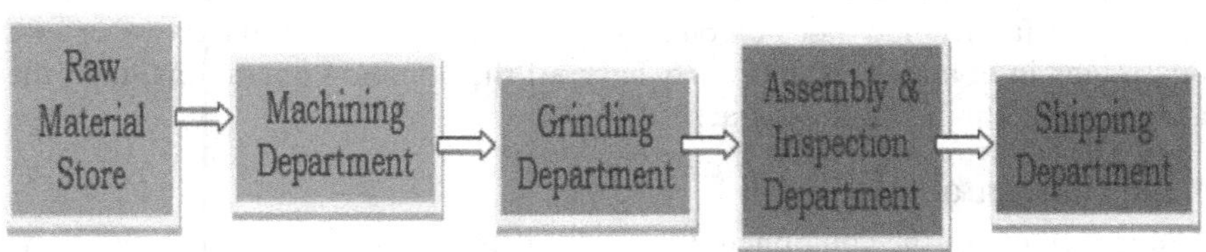


**FIGURE 7.1** Traditional manufacturing layout

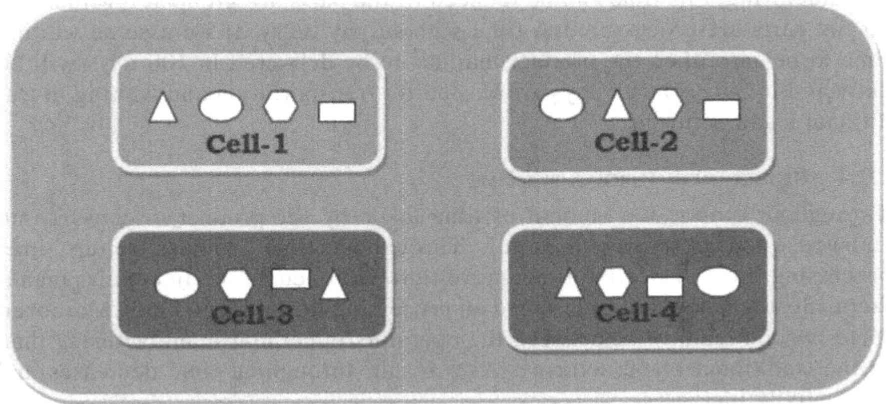


**FIGURE 7.2** Cellular manufacturing layout

when compared to traditional layout. The movement takes place only between the machines, and similar parts are divided and operated in their respective cells.

### 7.3.2 Reducing Set-up Time

Set-up time is nothing but the time taken to fix the component in the machine. Consider, for example, an operation on lathe. Figure 7.3 shows the cylindrical and cube-shaped components. If we consider three components of each time, the set-up is to be done three times one after the other in the traditional manufacturing using a three-jaw and four-jaw chucks, respectively, in lathe. But in cellular manufacturing the set-up is done once by grouping all cylindrical components on one cell and cube-shaped components on the other cell. Thus, 75% of time is saved.

### 7.3.3 Reduction in Waiting Time

The amount of time in which the component is waiting to be processed on a machine is known as waiting time. As per study conducted by researchers, 93% of manufacturing time is lost due to the waiting of product. In traditional manufacturing, this waiting time leads to decrease in productivity. In cellular

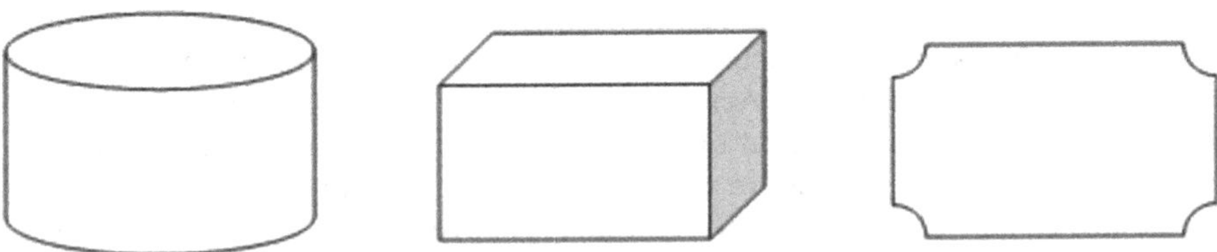

**FIGURE 7.3** Sample parts used for manufacturing before set-up

manufacturing, this time can be reduced as the parts are grouped together and all the parts are processed at a time without any delay. If we assume waiting time to be zero, then the product that has to be delivered in 100 days will be delivered in 5 days as 95 days are wasted for transportation and waiting in traditional manufacturing.

### 7.3.4 Reduction in Throughput Time

Throughput time is the amount of time taken by the product to convert into finished good from raw material. Throughput time includes set-up time, processing time, inspection time, move time and queue time. In cellular manufacturing, the amount of time spend on processing the product is less. Moreover, there is good team work between the operators in cellular manufacturing than in the traditional manufacturing. As a result, throughput time decreases and productivity increases.

## 7.4 LITERATURE REVIEW

In the recent years, there has been increasing application of artificial intelligence and neural network to the optimization problems. Many researchers driven the usage of those methods into the cell formation problems for machine part clustering and one such method with neural network framework is observed to solve the machine part groups [1]. The implementation of competitive learning by the neural network approach has addressed by Chu [2]. The implementation of pure neural networks based on the Capenter-Grossberg network has studied for formation of machines and parts into cells [3]. A dynamic scheduling method for the cell formation problems is based on artificial intelligence for the networking environment, by taking into account information such as loading factor, unexpected breakdowns and new job arrivals. Scheduling on both levels can be dynamically executed in real time [4]. A modified Hopfield neural network for cell formation problem, the quantized and fluctuated Hopfield neural network, was used by Ateme-Nguema and Dao [5] to solve the cell formation for big-size industrial dataset and obtain optimal solution. The fuzzy adaptive resonance theory (ART) neural network has been applied to solve the group manufacturing problems has been addressed and with the proposed Fuzzy ART based on a similarity measure, has a greater advantage over other existing approaches. Results of fuzzy ART applied to test different problems of part family formation and have given systematically generating alternative solutions in the cell formation problem [6]. A neural network approach based on a competitive learning paradigm was proposed to handle a cell formation problem. Computational experience shows that the procedure is fairly efficient and can effectively obtain optimal clustering results [2]. Artificial intelligence models, such as pattern recognition, have been used based on the patterns mined from the existing benchmark problems. This procedure involves no mathematical structuring, but due to the heavy logical computations for the large-sized datasets, the time taken is higher; however, the optimum solutions gained by both the GRASP [7] and the inspection-based clustering (IBC) algorithm are similar.

## 7.5 INSPECTION-BASED CLUSTERING ALGORITHM

The IBC algorithm has five basic elements to operate on and the following are the detailed list. The detailed layout is shown in Figure 7.4. In this chapter, the first three elements will be explained manually.

- Raw data component.
- Dynamic I/O table.
- Cognition panel.
- Output data component.
- Time evaluator.

### 7.5.1 Raw Data Component

In Figure 7.4, the first portion, i.e. the raw data, is provided by the user and it comprises binary data (0's and 1's), where 1 implies the corresponding machine component operation and 0 implies no operation. This matrix is said to be machine component incidence matrix.

### 7.5.2 Dynamic I/O Table

In Figure 7.5, we represented machine component incidence matrix with 1's and 0's and then we formed the I/O (in/out) table, which is nothing but 1's represent component is processed and 0's represent component is idle under that machines. The first row of an I/O table indicates the input, intermediate and output machines of a component 1. Here, we see that component 1 has machine 1 as input and machine 6 as output and machines 2 and 4 act as intermediate machines for component 1. Similarly, all the components, input machines, intermediate machines and output machines, are collectively called I/O table of component and this I/O table can also be formed for the machines called I/O table of machines. These two I/O tables can be generated for a given machine component incidence matrix.

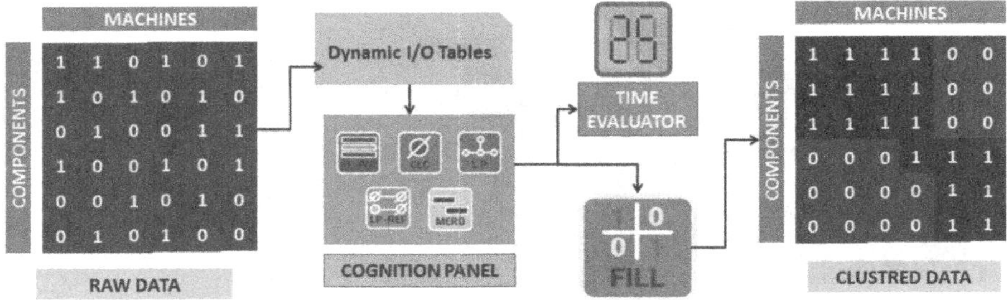


**FIGURE 7.4** Inspection-based clustering algorithm layout.

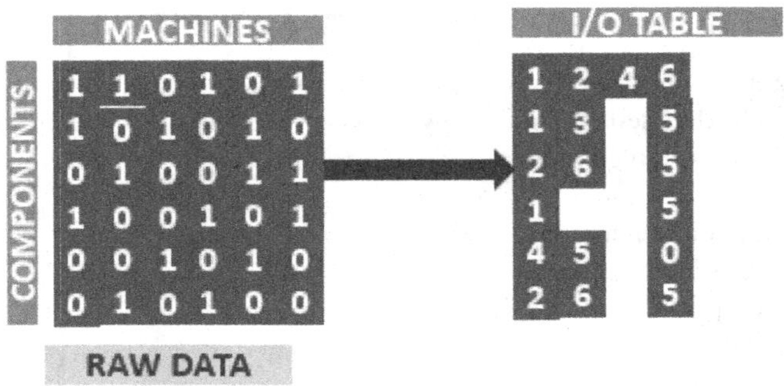


**FIGURE 7.5** Generation of I/O tables.

### 7.5.3 Cognition Panel

#### 7.5.3.1 Longest Path Module (LGP)

This cognitive algorithm checks for the longest node (by number of active machines) by priority and cancels them out to fill the data grid shown in Figure 7.6. It does not check for the next LGP if elements are exhausted, thereby cutting short the search by half. Supplemented by the cancellation module which is an integral update of data grid in the methodology.

Thus, after the generation of I/O table, component 1 has a greater number of operations (path is more) at various machines (1, 2, 4, 6) than other components (2, 3, 4, 5, 6), which have only two to three operations on machines. Hence, priority is given to the first component, which has 1 2 4 6 component sequence. Thus the sequence obtained by applying the longest path module is 1 2 4 6 components and so on this can method can even be done for machine I/O table too. For the complete sequence of machine or component can be obtained by repeatedly doing the same module.

#### 7.5.3.2 Similar Pattern Module

This cognitive algorithm checks for similar patterns or trends in data. Exhaustive search is of finite array and auto detection is based on dynamic array. May be supplemented by the cancellation module which is an integral update of data grid in the methodology.

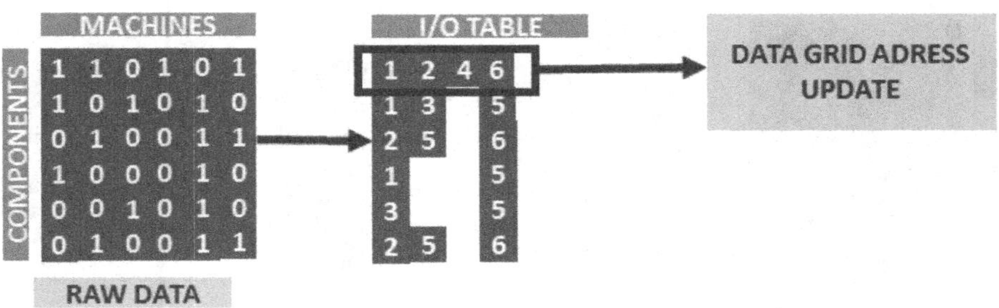


**FIGURE 7.6** Sequence from I/O tables using longest path module.

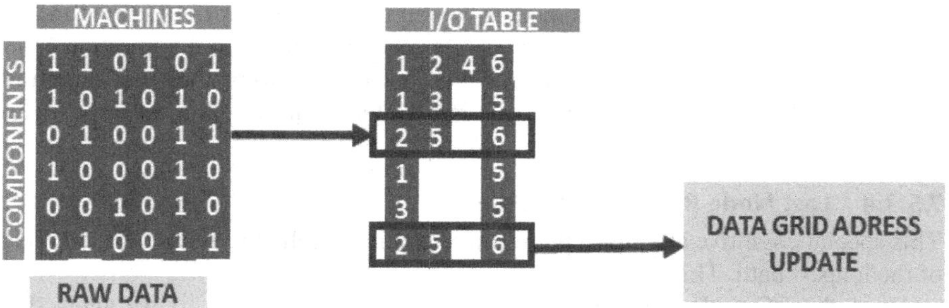


**FIGURE 7.7** Sequence from I/O tables using similar pattern module.

It can be seen from Figure 7.7 that component 3 and 6 have similar operation that are performed on the same machines (2,5,6). The I/O table for component 3 and component 6 is similar; hence, when we can apply this similar pattern module, the algorithm searches for the similar I/O's for any component or machines. If found similar, then the data grid is updated and sequence becomes 256 and so on since we are searching similar I/O's for components or machines. Hence, this algorithm is named as similar pattern module.

### 7.5.3.3 Cancellation Module

This cognitive algorithm cancels out the nodes in priority order. Lower search memory is a benefit of this method. After the I/O table formation if we apply this module then the cancellation phenomenon takes place, that is it cancels the starting machine number of the I/O table. If the cancelled machine is shown, it gives priority to the next machine number and sequence is obtained. This requires less time when compared to the other module because we are able to form the machine sequence or part sequence in less time. The priority will be in the order of natural number order (1, 2, 3, 4, 5, ...). This can be applied to machine I/O tables also, as shown in Figure 7.8. Component 1 is undergoing operations on 1, 2, 4, 6 machines and component 2 is undergoing operation on 1, 3, 5.

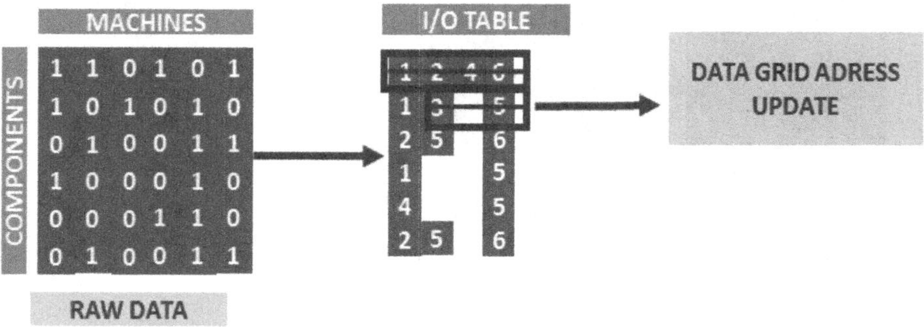


**FIGURE 7.8** Sequence from I/O tables using cancellation module.

Hence, after application of this module, we get the sequence as 1246 in component 1, since component 2 has machine 1 which is cancelled earlier in the machine 1, so the next priority is given to 3 and 5. Thus, on combination the final machine sequence obtained is (124635) and this is updated from the data grid shown in Figure 7.8.

#### 7.5.3.4 Last Node Reference Module

This special cognitive algorithm points out the non-cancelled reference valued node of the longest path. This module is case sensitive.

From the I/O table obtained for the machine or components, if we apply this module then the algorithm checks whether the last node is same for the other component's in or out machine. If found, it attaches the sequence of the child component to the parent component (initial components whose last nodes are matched) sequence. The last node may be in or out machine of any component. This can be applied to even machine's I/O table for sequencing. From Figure 7.9, component 1 has the machines as 1, 2, 4, 6. Here, the out machine of component 1 matches with the last machine of component 3, so component 3 takes the sequence as 6, 5 but not machine 2 because it is already taken in the component 1 sequence and finally this sequence will be updated to the initial component 1 sequence. Hence, the final sequence in the data grid will be updated as 12465.

#### 7.5.3.5 Merge Module

This special cognitive algorithm checks for the in and out nodes of different I/O array and it merges out the intermediate nodes in the path can be used with repeated patterns module which is bit more advance in implementation.

This is the phenomenon of merging two component sequences. After the formation of the I/O tables, if we apply this merge module, then the algorithm checks for the same input and output machines but with slightly more machine intermediately located on the other component. If such patterns are found, it uses the merge phenomenon in which two component sequences are merged to form the final sequence. By

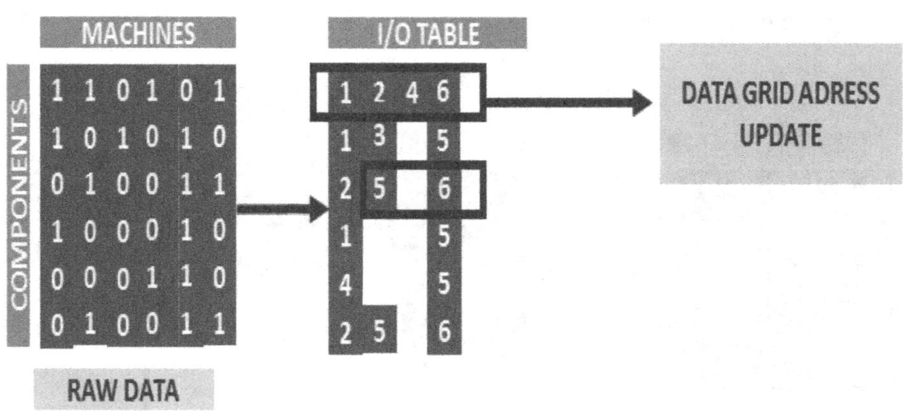


**FIGURE 7.9** Sequence from I/O tables using last node reference module.

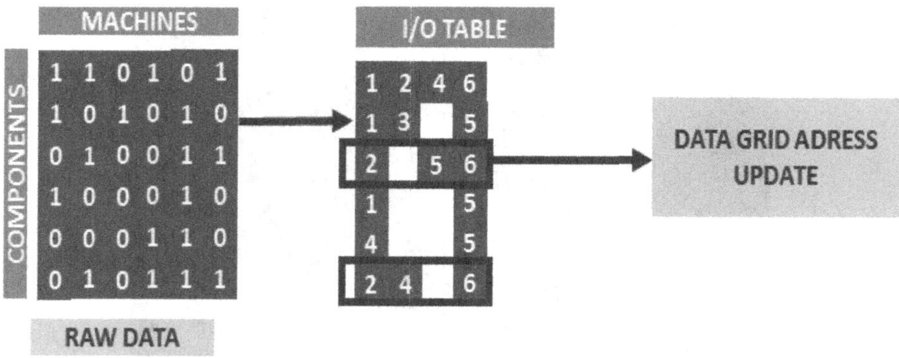


**FIGURE 7.10** Sequence from I/O tables using merge module.

combining the intermediate machines, this can be applied to machine I/O table too. In Figure 7.10, component 3 has the machines as 2,5,6 and the same input and output machines are similar for the component 6 with an extra intermediate machine of 4. So, this fourth machine will be updated to the initial three component sequences and gives the final updated sequence as 2456 on the data grid. The complete sequence can be formed by other modules application after merge module.

## 7.6 SOFTWARE MODULES DEVELOPED FOR INSPECTION-BASED CLUSTERING ALGORITHM

### 7.6.1 Machine Component Size

In this portion, we need to provide initially the size of the machine component incident matrix. For example, the problem is expressed as (5M × 6C), so we should give the number of rows or number of machines as 5 and the number of columns or number of parts as 6, as specified in the software. We can change the index, either numbers or alphabets, as shown below based on our choice. The input is given in the portion shown below in Figure 7.11 in the software.

### 7.6.2 Input Matrix View

After the specification of the size of the problem by clicking on the 'Generate Input Matrix', as shown in Figure 7.11, we can see the generated input matrix of given size for the data entry of our problem in the form of 0's and 1's. Here, we are showing

**FIGURE 7.11** Size of matrix is provided in this portion of software.

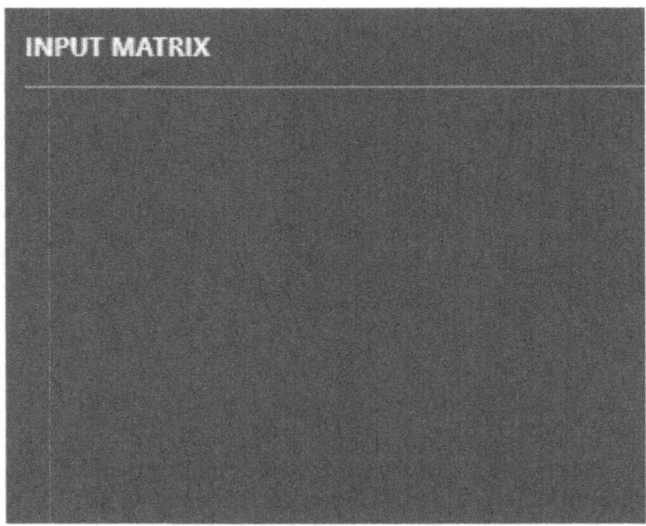


**FIGURE 7.12** Input matrix where the binary data is provided.

only the place where the input matrix is created but the actual created matrix for different size is shown and explained in detail in the below problems. Figure 7.12 shows the input matrix which will be formed by our specified size of the problem.

### 7.6.3 I/O Table Generation

Here, the input-output tables can be generated by the software which has a greater advantage of reducing the time required for the generation of I/O tables manually. The I/O tables can be generated and viewed for both the machines and components or parts individually by clicking on the options given in this module. After viewing, this also suggests the modules which can be applied. The module gives the options like generation of I/O table and viewing I/O tables and viewing input matrix for the rechecking. The figure of the module is shown in Figure 7.13.

### 7.6.4 Machine and Part Table Methods

As discussed earlier, we have five modules, namely longest path, merge, cancellation, last node reference and similar pattern techniques. All these techniques are combined and shown in this module for the part and machine sequencing shown in

**FIGURE 7.13** Machine and part tables can be generated using the option.

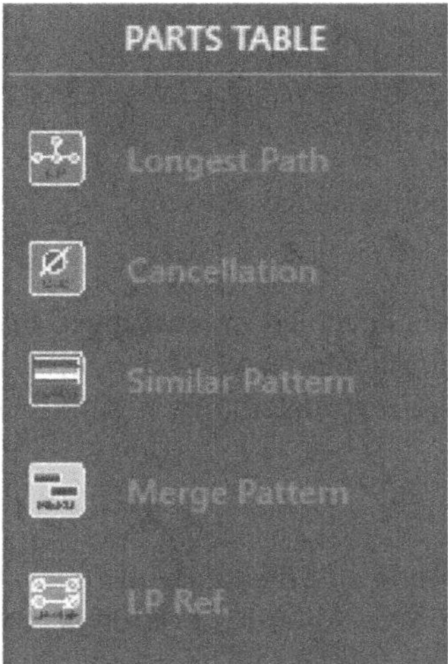


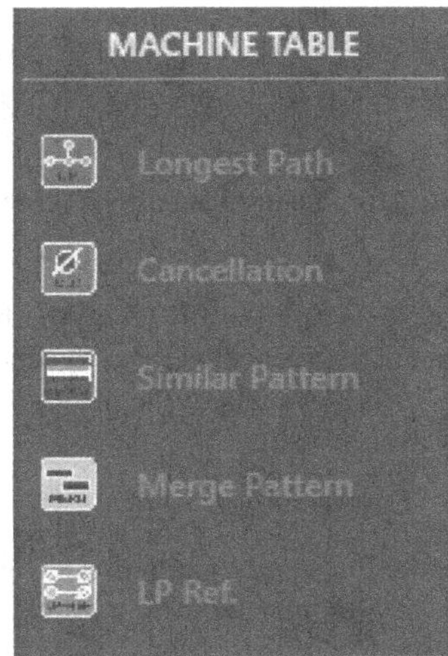


**FIGURE 7.14** Machine and part table modules to form the sequence for cells.

Figure 7.14. We can use different techniques individually or any combination of the techniques based on our problem. We have shown the I/O table generation where it also suggests the type of techniques that can be used for a particular given problem. In fact, the software suggests you the application of particular technique for this problem. Therefore, we can sequence the parts or machines just by clicking on the particular technique. In this module, the sequenced parts and machine are shown on the preview table which will be explained in the preview module.

### 7.6.5 Preview Table

After the application of different modules for both machine and components, the sequenced parts and machines are obtained. Machines' and components' matrix is filled back with the data given in the input data (1's or 0's) for machine component matrix. It is nothing but filling back the data by using the fill option based on the obtained sequenced matrix at the output. By this we can get the final block diagonal form where we can form cells, machine cells, part families, exceptional components and exceptional elements later. This is obtained by using the option shown in Figure 7.15.

**FIGURE 7.15** Rearranged matrix is obtained by selecting fill table.

FIGURE 7.16 Rearranged matrix is viewed in this portion of the software.

After selecting the fill table, the binary data is occupied in rearranged machine component incidence matrix by application of Machine and Part table's. Below Figure 7.16 shows where the preview table or final rearranged matrix is formed.

### 7.6.6 Elapsed Time

This is the major module in this software which indicates the total time taken by the CPU of a computer to solve a problem. The time viewed in this portion is generally represented in milliseconds (ms). The time taken by the CPU for solving will be different for different problems, and it depends on the data and the size of the initial machine component incidence matrix. The time for a particular problem is varied based on the configuration of the system depending on its RAM and processor. The elapsed time in milliseconds will be displayed as shown in Figure 7.17.

### 7.6.7 Software Window with All Modules

All developed modules in IBC interface are shown below in Figure 7.18.

## 7.7 APPLICATION OF ALGORITHMS FOR CELL FORMATION PROBLEMS

Cellular processing uses cell-forming group technology. Grouping technology's main objective is to form part families based on similar processing requirements. Based on similar manufacturing process techniques, parts and machines are grouped. This approach results in cells where machines are very close to each other based on the criteria for storage rather than different functional aspects. Decision-making and transparency are more centred locally, often contributing to increases in reliability and profitability.

Several cell formation techniques came into existence; among those, we considered rank order clustering (ROC) [8] and GRASP [7] to compare with IBC algorithm. Performance measures play a key role in determining the optimum solutions

FIGURE 7.17 Time can be noted in milliseconds from this portion in software.

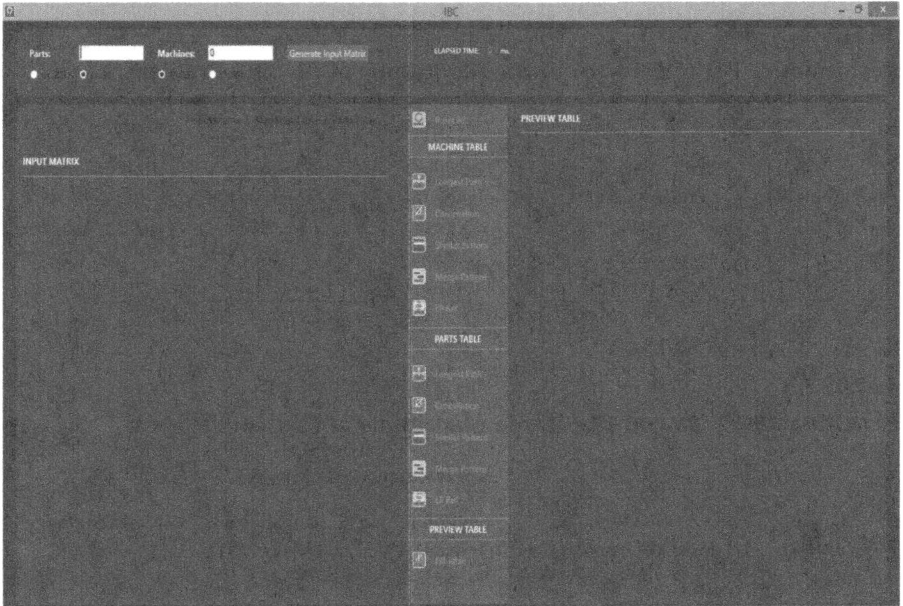


**FIGURE 7.18** IBC interface with the five modules was implemented using software.

for the obtained cells. Various measures considered are machine utilization, grouping efficiency, grouping efficacy and cell efficiency.

A benchmark problem set of size (7 × 11) collected from Boctor [9] is solved by using ROC, GRASP and IBC modules. The various performance measures are calculated and shown in Table 7.1.

A benchmark problem set of size (8 × 20) collected from Chandrasekharan and Rajagopalan [10] is solved by using ROC, GRASP and IBC modules. The various performance measures are calculated and shown in Table 7.2.

A benchmark problem set of size (5 × 18) collected from Seifoddini [11] is solved by using ROC, GRASP and IBC modules. The various performance measures are calculated and shown in Table 7.3.

**TABLE 7.1**
**Performance Parameters for Three Algorithms of (7 × 11) Problem**

| | Performance Measures | | | |
|---|---|---|---|---|
| **Algorithms** | **Machine Utilization (MU)** | **Grouping Efficiency (Gr.E)** | **Grouping Efficacy (T)** | **Cell Efficiency (η)** |
| GRASP | 0.760 | 0.841 | 0.703 | 0.817 |
| ROC | 0.760 | 0.841 | 0.703 | 0.817 |
| IBC | 0.760 | 0.841 | 0.703 | 0.817 |

**TABLE 7.2**
**Performance Parameters for Three Algorithms of (8 × 20) Problem**

| Algorithms | Performance Parameters | | | |
|---|---|---|---|---|
| | Machine Utilization (MU) | Grouping Efficiency (Gr.E) | Grouping Efficacy (T) | Cell Efficiency (η ) |
| GRASP | 0.958 | 0.836 | 0.696 | 0.712 |
| ROC | 0.958 | 0.836 | 0.696 | 0.712 |
| IBC | 0.98 | 0.871 | 0.777 | 0.782 |

**TABLE 7.3**
**Performance Parameters for Three Algorithms of (5 × 18) Problem**

| Algorithms | Performance Parameters | | | |
|---|---|---|---|---|
| | Machine Utilization (MU) | Grouping Efficiency (Gr.E) | Grouping Efficacy (T) | Cell Efficiency (η ) |
| GRASP | 0.823 | 0.800 | 0.736 | 0.755 |
| ROC | 0.854 | 0.808 | 0.773 | 0.768 |
| IBC | 0.854 | 0.808 | 0.773 | 0.768 |

## 7.8 RESULTS AND DISCUSSION

- In the graph shown in Figure 7.19, if we see the performance parameters of a cell formation problem such as machine utilization, grouping efficiency, grouping efficacy and cell efficiency of the standard problem set [9] obtained by traditional ROC technique, the actual optimum solution obtained in the published paper on GRASP technique and our new algorithm, inspection-based clustering, is same. This means that this algorithm is one of the best techniques for obtaining actual and exact solution.

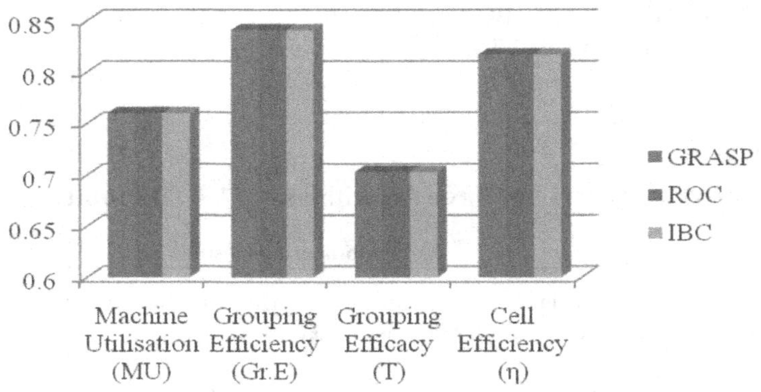


**FIGURE 7.19** Comparison of performance parameters calculated in Table 7.1.

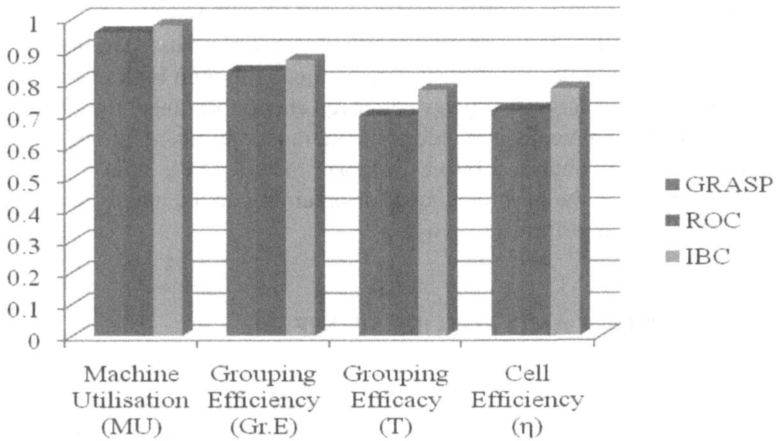


**FIGURE 7.20** Comparison of performance parameters calculated in Table 7.2.

- In the graph shown in Figure 7.20, we see the performance parameters of a cell formation problem such as machine utilization, grouping efficiency, grouping efficacy and cell efficiency of the standard problem set [10] obtained by traditional rank ordering clustering technique and the actual optimum solution obtained in the published paper on GRASP technique are same but our new algorithm, inspection-based clustering, gives better result than the other two techniques, which means it is able generate most optimal solutions than the actual solutions obtained by best algorithms till now (ROC and GRASP). This means that this algorithm is the best technique than the other two techniques for obtaining actual and exact solution.
- In the graph shown in Figure 7.21, if we see the performance parameters of a cell formation problem such as machine utilization, grouping efficiency, grouping efficacy and cell efficiency of the standard problem set [11] obtained

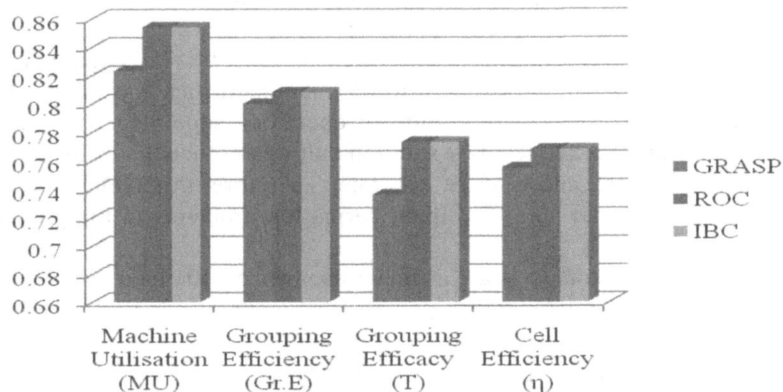


**FIGURE 7.21** Comparison of performance parameters calculated in Table 7.3.

by our new algorithm are more than the GRASP technique, which gave less values compared with our new algorithm inspection based clustering and traditional rank ordering clustering technique, although both of these give most optimum solutions. It gives better result (performance) than the other technique, which means it is able generate most optimal solutions than the actual solutions obtained by best algorithm (GRASP). This means that this algorithm is the best technique for obtaining actual and exact solution than the other algorithms.

## 7.9 CONCLUSION AND FUTURE SCOPE

### 7.9.1 Conclusion

The development of the IBC has helped in solving problems with ease with no mathematical modelling, but with successful implementation of the partial artificial intelligence module (Pattern recognition) has been used which is the novel idea to prove that all the cellular or the binary related problems can be solved using IBC. The performance of the benchmark problem sets, as discussed in results and discussion section, has proved that the IBC and its internal algorithms can help in solving large dataset algorithms in minor time span. We have developed the software for the IBC algorithm that gives more feasible output, is accurate with less efforts and involves less computations with less processing time. We can get the solutions for any type/size of machine component incident matrix. Thus, we can say that this will be one of the best algorithms to solve cell formation problems.

### 7.9.2 Future Scope

The present study only discussed few patterns mined from the benchmark problems. More patterns have to be studied and need to enclose in the tool or such studies can be extended. These methods can be implemented in the field of the organization management and transportation optimization problems.

Implementation of cellular manufacturing is not just by rearranging of facilities, which is of interest and concern to the manufacturing floor alone. It also brings changes that affect work culture of operators, supervisor and designed departments. Thus, we can generate solution that will reduce the investment cost and requires less time and is more accurate. Therefore, this technique can be applied in practical and natural cellular formation problems in the production industry.

Since the formulated models are artificial intelligence inherited and the formulations speak about the optimization process, this method suits well. The derived models can help in solving the travelling sales man problem in different perspective by natural maximizing function based on the maximum number of the nodes in a cluster. Natural controlling and data rephrasing can be easily evaluated and further helps in the determining the risk management and analysis.

## REFERENCES


1. Y. B. Moon, "Forming part-machine families for cellular manufacturing: A neural-network approach", *The International Journal of Advanced Manufacturing Technology,* Volume 5, Issue 4, 1990, pp. 278–291.
2. Chao Hsien Chu, "Manufacturing cell formation by competitive learning", International *Journal of Production Research,* Volume 31, Issue 4, 1993, pp. 829–843.
3. Shashidhar Kaparthi and Nallan C. Suresh, "Machine-component cell formation in group technology: a neural network approach", *International Journal of Production Research,* Volume 30, Issue 6, 1992, pp. 1353–1367.
4. Michael J. Shaw, "Dynamic scheduling in cellular manufacturing systems: A framework for networked decision making", *Journal of Manufacturing Systems,* Volume 7, Issue 2, 1988, pp. 83–94.
5. Barthelemy Ateme-Nguema and Thien-My Dao, "Quantized Hopfield networks and tabu search for manufacturing cell formation problems", *International Journal of Production Economics,* Volume 121, Issue 1, 2009, pp. 88–98.
6. Laura Burke and Soheyla Kamal, "Neural networks and the part family/machine group formation problem in cellular manufacturing: A framework using fuzzy ART", *Journal of Manufacturing Systems,* Volume 14, Issue 3, 1995, pp. 148–159.
7. Juan A. Diaz, Dolores Luna and Ricardo Luna, "A GRASP heuristic for the manufacturing cell formation problem", *TOP* (An Official Journal of the Spanish Society of Statistics and Operations Research), Volume 20, Issue 3, 2012, pp. 679–706.
8. J. R. King, "Machine-component grouping in production flow analysis: An approach using a rank order clustering algorithm", *International Journal of Production Research,* Volume 18, Issue 2, 1980, pp. 213–232.
9. Fayez F. Boctor, "A linear formulation of the machine-part cell formation problem", *International Journal of Production Research,* Volume 29, Issue 2, 1991, pp. 343–356.
10. M. P. Chandrasekharan and R. Rajagopalan, "An ideal seed non-hierarchical clustering algorithm for cellular manufacturing", *International Journal of Production Research,* Volume 24, Issue 2, 1986, pp. 451–464.
11. Hamid K. Seifoddini, "Single linkage versus average linkage clustering in machine cells formation applications", *Computers and Industrial Engineering,* Volume 16, Issue 3, 1989, pp. 419–426.
12. B. S. Nagendra Parashar, "Cellular manufacturing systems: An integrated approach", PHI Learning Private Limited, 2009.
13. Yong Yin and Kazuhiko Yasuda, "Similarity coefficient methods applied to the cell formation problem: A comparative investigation", *Computers and Industrial Engineering,* Volume 48, 2005, pp. 471–489.
14. K. V. Durga Rajesh and P. V. Chalapathi, "An efficient sheep flock heredity algorithm for the cell formation problem", *ARPN Journal of Engineering and Applied Sciences,* Volume 12, Issue 21, 2017, pp. 6074–6079.
15. K. V. Durga Rajesh and P. V. Chalapathi, "Application of efficient SFHA and TLBO algorithms for cell formation problems in cellular manufacturing environment", *International Journal of Mechanical and Production Engineering Research and Development,* Volume 9, Issue 3, 2019, pp. 43–52.
16. K. V. Durga Rajesh and P. V. Chalapathi, "Performance analysis of enhanced cell formation techniques in a manufacturing industry – A case study", *International Journal of Innovative Technology and Exploring Engineering,* Volume 8, Issue 6, 2019, pp. 574–578.
17. K. V. Durga Rajesh, "MATLAB based Manhattan distance matrix method to solve cell formation problems", *International Journal of Innovative Technology and Exploring Engineering,* Volume 8, Issue 8, 2019, pp. 3102–3105.

# 8 Optimization of Operating Parameters of Wire EDM

*Shaikh Zubair A.*
Shreeyash College of Engineering & Technology
Aurangabad, Maharashtra, India

*Swarup S. Deshmukh*
National Institute of Technology
Durgapur, West Bengal, India

*Dheeraj Kumar*
National Institute of Technology
Durgapur, West Bengal, India

*Vijay S. Jadhav*
Government College of Engineering
Karad, Maharashtra, India

*Ramakant Shrivastava*
Government College of Engineering
Karad, Maharashtra, India

**CONTENTS**

## 8.1 INTRODUCTION

The requirement of alloy materials with great hardness, strength and impact tolerance has been raised by the growth of the automotive sector to satisfy diverse needs. However, such materials are difficult to machine using conventional machining techniques. Therefore, non-traditional machining techniques such as electrochemical, ultrasonic and electrical discharge machining (EDM) are employed to machine materials that are too difficult to machine. As an electrode, the thin wire EDM (WEDM) method converts electrical into thermal energy for cutting materials. Composite materials, electroconductive ceramic and aircraft materials can be machined regardless of their hardness and durability through this process. In addition, WEDM is able to create a fine, detailed surface that is resistant to corrosion and wear [1].

WEDM, which utilizes a wire electrode to initiate the sparking mechanism, is a special form of the traditional EDM process. However, WEDM uses a constantly moving wire electrode of 0.05–0.30 mm in diameter of thin copper, brass or tungsten, which is responsible for producing very limited corner radii. Using a mechanical tensioning system, the wire is held under tension, reducing the propensity to generate incorrect sections. The steel is eroded in front of the wire during the WEDM process, and there is no immediate interaction between the working piece and the wire, reducing the mechanical stresses during machining [2].

WEDM is a method of thermal electric spark erosion to cut extremely tough conducting material with the aid of a wire electrode continually feed through the workpiece that allows high precision machining of part of complex shapes. The wire is wound and is held at steady tension on a spool. Fresh wire is constantly supplied to the work area by the drive mechanism. During the machining, the substance to be removed is sprayed with deionized water. A number of separate sparks that occur within the workpiece and the wire electrode erode the material from the work material. The WEDM system uses electrical energy to create a plasma channel between the wire electrode and the workpiece and transform it into heat energy of about 8,000–12,000°C, initializing a large amount of material on the workpiece surface to melt and vaporize. The positioning system to produce the correct shape to the workpiece is controlled by computer-based numerical control via the input program provided. It is quite commonly used for pattern and die-making industries in hard materials. The technique is often used to cut complicated shapes for parts used in the electrical and aerospace industries [3].

## 8.2 PAST WORK AND SCOPE OF CURRENT WORK

### 8.2.1 Literature Review (Past Work)

Ugrasen et al. [4] employed orthogonal matrix ($L_{27}$) for the creation of a rigid model when machining SS304 steel. From this study, they determined that the pulse-on time ($T_{on}$) has more influence on surface roughness (SR) of the components, that the current has more impact on the precision and that the speed of the bed has much more effect on the material removal rate (MRR). Kumar et al. [5] examined surface alteration of tungsten carbide–cobalt alloy using aluminium and silicon powder of 400 mesh scale during WEDM. Goswami and Kumar [6] conducted trim cutting machining and surface characterization of Nimonic-80A alloys through WEDM. The outcome of this analysis is that $T_{on}$ has a significant impact factor for MRR and SR and the wire wear ratio is primarily influenced by the wire offset factor. Sample machined at high-energy input conditions provides rugged surface finish with a number of built-up edges, while a decent surface finish is accomplished at lower energy input.

Goyal [7] studied MRR and SR with cryogenically treated wire electrode during WEDMs of Inconel 625 super alloy. The finding of this experiment is that the cryogenically processed electrode offers a better result for MRR and SR. The key determining parameters are current and pulse-on time for cryogenic treated electrode. Dabade and Karidkar [8] implemented the Taguchi $L_{27}$ orthogonal array while machining Inconel 718 to build a robust configuration. Finally, they concluded that, among the other parameters, rank table $T_{on}$ has a greater effect on MRR, SR and kerf distance. Tilekar et al. [10] used the Taguchi approach to perform operating parameter optimization of WEDM on aluminium and mild steel. They concluded that spark time has more influence on SR of aluminium and current has more impact on SR of mild steel. On the other side, wire feeding and spark time have more impact on kerf distance during WEDMs of aluminium and mild steel. Mahapatra and Patnaik [11] implemented the Taguchi approach to optimize the operating parameters of WEDM, and for multi-target optimization, they employed a genetic algorithm. The experiment concluded that the errors linked to MRR, surface finish and kerf distance were 3.24, 1.55 and 3.53%, respectively. Chiang and Chang [12] performed WEDM process optimization of particle-strengthened material using grey relationship analysis based on Taguchi. $Al_2O_3$ particle strengthened material (6061) alloy was used as workpiece and 0.20-mm-diameter copper wire was used as an electrode in the experimentation. The result of this experiment is that the maximum SR is decreased to 3.214 to 2.051 μm for single output characteristics; the MRR is raised from 7.203 to 14.102 mm$^2$/min. The SR and MRR after multi-objective optimization are 2.305 μm and 14.616 mm$^2$/min.

### 8.2.2 Scope of Current Work

In the fast-growing technological world, every manufacturing company wants economic and fast way of machining. To address the need for accuracy, reduced surface finish and complex geometry, WEDM has evolved in the last three decades. The only problem with WEDM is its cutting speed. As it cuts intricate shapes and complex geometry with wire, it takes longer time. To cope up with this problem, electrical conductivity of wire or workpiece is to be increased. Because in the WEDM, the cutting takes place due to thermoelectric spark, this thermoelectric spark causes the erosion of material. In order to increase the thermal and electrical conductivity, cryogenic treatment. In order to increase the thermal and electrical conductivity, cryogenic treatment of wire and workpiece is to be done. By doing so, cutting speed can be increased. AISI D2 steel finds numerous applications in die- and punch-making industry. Cryogenic-treated D2 steel and cryogenic-treated Bronco cut wire are never being used for the machining as per literature survey. Investigation of optimum parameters for cryogenic-treated and normal wire and workpiece is essential for comparing the results and checking the scope of cryogenic treatment in WEDM. Taguchi-based optimization will be used in this work for finding optimum process parameters for machining of D2 steel.

### 8.2.3 Basics of Cryogenic Process

Generally, the method of thermal treatment has been utilized to enhance the hardness of cutting tools, while cryogenic process is utilized as an auxiliary treatment to raise the hardness and wear properties. However, the literature review indicates that thermal treatment accompanied by cryogenic treatment is useful for attaining the intended properties and increasing efficiency. The schematic of cryogenic treatment is shown in Figure 8.1.

Cryogenics, a low-temperature physics branch, focuses on the impact of extremely low temperatures below around 123°K (–150°C) and stretches to absolute zero (–273°C). Its application temperature, below 123°K or at around liquid nitrogen temperature (–196°C), will describe the cryogenic care. Cold therapy requires temperatures below zero but higher than cryogenic (~ –80°C) temperatures. They consist of three major phases, including: the slow-cooling process in which the components

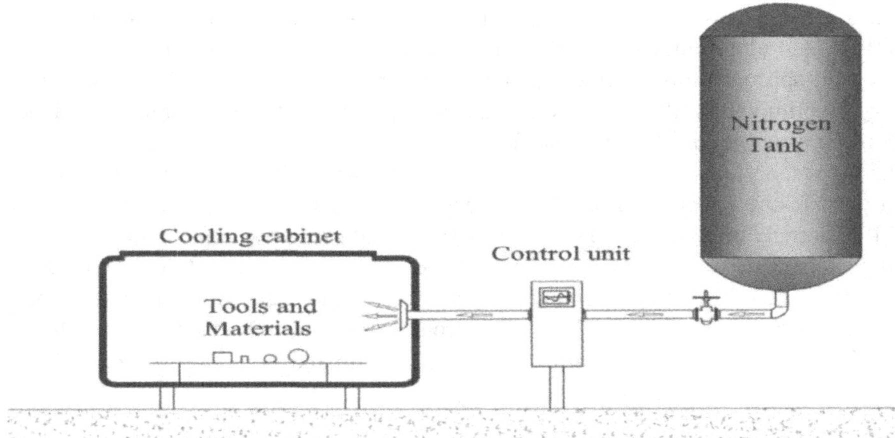


**FIGURE 8.1** Schematic of cryogenic treatment.

are cooled from room temperature to cryogenic temperature for a period of time (degrees/hour or minute); the soaking phase in which the components are retained for a given time period at cryogenic temperature; and the warming phase (warm-up cycle/period) in which the components are heated from cryogenic temperatures [13].

AISI D2 tool steel's cryogenic treatment affects its micro-structural properties, which makes it stiff and durable and therefore increases its wear resistance. It has been stated that D2 steel requires 36 hours of soaking time to achieve the required micro-hardness and SR. Owing to the transmutation of preserved austenite into martensite, the hardness of a deeply cryogenically processed sample is improved by 1.26%. The surface finish of deeply cryogenically processed AISI D2 steel alloys samples increased by 13.43%. In deep cryogenic therapy, the enhanced surface finish is often taken as an indicator of the decreased distortion [14].

The goal of this analysis is to examine the impact of parameters for WEDM on SR. In addition to the Bronco cut wire (beta brass-coated copper), D2 tool steel (cryogenic-treated) is used. In the die and punch industries, where wear occurs due to repeated die-punch operations, D2 steel is used. The wear resistance and lifespan of the product can be improved by cryogenic treatment of workpiece material and wire electrodes.

### 8.2.4 Stages of Cryogenic Treatment

- **Rate of cooling:** In cryogenic heat therapy, there are two key techniques for the rate of cooling. The first solution is called thermal shock, where the samples are quickly taken to the temperature of a cryogenic application. The samples are steadily taken to cryogenic temperature in the other method. An analysis of previous research shows that micro-cracks form on the specimens when the specimens are exposed directly to cryogenic heat because of the rapid increases in temperature, resulting from thermal shock.
- **Soaking time:** In the transition of phase from austenite to martensite, the development of novel carbides and the distribution of carbide, soaking time is an important parameter. To achieve optimum productivity and minimize costs for cryogenic treatment, optimizing the soaking time is critical. An analysis of the literature on the cryogenically processed cutting tools manufactured from austenitic materials shows that they are mostly subjected to the soaking duration of between 1 and 40 hours as well as 9 hrs, 24 hrs to examine the wear and corrosion resistance, toughness, etc. They suggested that a 9-hour soaking time for wear conduct was sufficient. Samples were subjected to 12-, 24- and 36-hour soaking cycles to give martensite stainless steel (SR34, containing 18% Cr) shows maximum wear resistance. With a soaking time of 36 hours, the researchers announced that they had obtained the best wear resistance. In comparison, the 36-hour soaking cycle resulted in a 24% improvement in the wear resistance of the specimens. Cryogenic therapy was applied to HSS (M2) cutting tools by maintaining them for 8, 16 and 24 hours. They recorded that a 24-hour held time had yielded improved results in cutting torque, SR and wear resistance in drilling experiments. During cryogenic therapy, the researchers estimated a 35% increase in workpiece material SR. The impact of cryogenic characteristics

over the mechanical characteristic of 18NiCrMo5 carbide steel was studied by Baldissera and Delprete [15]. The researchers measured the 1 hour (M1) and 24 hours (M24) soaking times of cryogenic treatment and concluded that in samples kept for a 24-hour soaking period, the stiffness and tension were greater than in those held for a 1-hour soaking period. The influence of cryogenic treatment factors on Ti-6Al-4V alloy was analyzed and the impact of various cryogenic process times on the hardness of samples were compared (at a 72-hours keeping time, the authors reached optimum hardness). In general, comparable advances have been accomplished in cryogenic process employed in cutting tools as compared to steel, according to the literature. At smaller or larger holding times, however, some experiments have obtained better results. Depending on the material of the cryogenic processed tool and the machining form of the tool, the results obtained with various soaking times differed.

- **Soaking temperature:** Depending on the kind of tool, cryogenic process performed at low temperatures reveals differences. Cryogenic treatment temperatures vary between –80 and –196°C for cutting equipment. The required soaking temperature should be calculated for the tool that will follow cryogenic process in order to improve material characteristics such as wear resistance, hardness and durability. In order to find the right one for cryogenic therapy, multiple experiments performed for this purpose have sought to accomplish optimization by varying soaking temperatures. Among the cryogenic process parameters, it has been stated that soaking temperature is the most powerful parameter in improving the samples' wear resistance. The optimal cryogenic process temperature has been calculated as –184°C by researchers. Previous experiments have shown that the results of various temperatures of cryogenic therapy vary depending on the type of material.
- **Method of tempering:** After cryogenic procedure, cutting tools are usually tempered. To remove inside stresses from the cutting tool that arise due to extreme cooling, the tempering process is performed. Generally, by keeping them at 150–200°C for 1.5–2 hours, the tempering process is extended to cutting instruments. In certain previous experiments, following cryogenic treatment, in excess of one tempering phase was implemented successively. The cryogenic process and tempering period's primary purpose is to improve hardness and transform the material from austenite phase to martensite phase. Another intention is to raise wear resistance by enhancing the microstructure's carbide dispersion [16].

## 8.3 EXPERIMENTAL SET-UP (WEDM MACHINE)

Experiments were carried out on electronic sprint-cut WEDM at Indo German Tool Room (IGTR), Aurangabad. The pictorial view of the experimental set-up is shown in Figure 8.2.

The material for cutting was AISI D2 steel (normal and cryogenic-treated) and Bronco cut wire (normal and cryogenic-treated). The sprint-cut WEDM machine comes with a machine tool, power supply section and dielectric unit which filters the water and also deionize it. There is also a (Computer Numerical Control) CNC

**FIGURE 8.2** Pictorial view of the experimental set-up.

unit which takes the input in the form of program to guide the wire for specific cutting of the workpiece. The wire from the spool is rounded from many pulleys and guided to upper and lower nozzle. The workpiece is mounted between upper and lower nozzles from which deionize water is flushed to clean the workpiece. The wire is then directed to the waste chamber.

## 8.3.1 Major Components of WEDM Machine

### 8.3.1.1 Physical Device

WEDM's physical machine consists of a coordinate worktable, wire feeding mechanism, control cabinet operated by a micro-computer, dielectric supply unit, collector of wire wastage and worktable controller. With the support of clamps and bolts, the work material is fixed on the worktable. With the help of variable gears, lead screws and bolts, the micro-controller provides the pulse signals to servomotors that rotate accordingly. To execute the cutting process, these movements will be transmitted to the worktable.

### 8.3.1.2 Worktable

It consists of a primary worktable (called the X–Y table) and a secondary table (called the U–V table). The workpiece is placed on the primary worktable and clamped. By way of servomotors, the main table travels across the X and Y axes, in steps of 1 μm. In steps of 1 μm, these servomotors can travel with precision. The servo systems are the worktable's muscles that allow it to travel across the programmed path. The worktable is managed by the CNC device and follows the inputs provided by a programmer.

### 8.3.1.3 Display Control

The micro-computer-based control cabinet is shown in Figure 8.2. This control cabinet is used to program, control and supply pulses and electrical components of machines that are assembled into a single CNC device.

#### 8.3.1.4 Wire Electrode

The wire electrode is the WEDM important tool which never makes contact with the machined material. The wire diameter ranges from 0.20 to 0.30 mm, but 0.25 mm is found to be the best. The wire originated from a spool and then passes through a tension system (to hold it straight, different diameter wire needs different tension). Then, where the current is supplied, it makes contact with power feed contacts. The wire then travels through the circular diamond guides, and is then wounded over the spool of waste wire. Since it is eroded by the EDM method, the wire can only be used once. From the time WEDM was first adopted the key concern was wire material, so this material should have physical characteristics such as conductivity and high tensile strength. It is necessary to have a good conductivity rating because it ensures that the wire can bear more current, which is equal to hotter spark and improved cutting speed, at least technically. The selection of wire essentially depends on the characteristics of the workpiece material, but the best wire electrode should have characteristics of high electrical conductivity, adequate tensile strength and optimal characteristics of spark and flushing.

### 8.3.2 Selection of Workpiece Material

AISI D2 steel is the material chosen for this work. It is steel alloyed with molybdenum and vanadium, a high carbon and high chromium cold work product. It has various characteristics, such as excessive wear resistance, maximum compressive strength and high hardening stability and strong back tempering resistance.

It is heat-curable and provides hardness in the 55–62 HRC range and can be machined in the annealed state. On proper hardening, D2 steel exhibits no distortion. The high chromium content of D2 steels makes it moderate corrosion-resistant material in the hardened state. AISI D2 steel is used to shallow drawing and shaping dies, hobbing, laminating and stamping dies, shear knives, blanking, slitting cutters, spinning thread and wire dies, extrusion dies, etc. In all sorts of industries, the special characteristics of this alloy have made it useful. Because of these qualities, it is complicated to machine this alloy using traditional methods. This can be used to produce dies and punches through WEDM. The chemical composition of AISI D2 steel is given in Table 8.1.

#### 8.3.2.1 Properties of AISI D2 Steel

Physical properties:

- Density: 7,696 kg/m$^3$.
- Modulus of elasticity: 207 GPa.
- Thermal conductivity: 41.5 W/m/K.
- Machinability: 55 to 65%.

**TABLE 8.1**
**Composition of Material**

| Element | C | Si | Mg | Cr | V | Mo | Fe |
|---|---|---|---|---|---|---|---|
| Weight % | 1.51 | 0.30 | 0.30 | 11.87 | 0.9 | 0.67 | Balanced |

- Hardness: 60 HRC.
- Specific gravity: 7.75.

To enhance the electrical conductivity and thermal conductivity, AISI D2 steel was cryogenic-treated at –196°C in this work.

### 8.3.3 Selection of Wire Electrode Material

There are various types of wire material which can be used according to applications. Generally used wire materials are:

- Plain wires (brass wire, copper wire, etc.).
- Coated wires (zinc-coated brass wire).
- Diffusion annealed wires (alpha brass, beta brass, etc.).
- Composite wires (steel core wire).

The choice of wire is usually dependent on the characteristics of workpiece material, but the optimal wire electrode should have characteristics of huge electrical conductivity, adequate tensile strength and optimal properties of spark and flushing. Beta brass-coated copper (Cu) wire will be chosen for this work. Owing to the larger percentage of Cu in it, Bronco cut wire has improved performance than other forms of wires. The cryogenic-treated and normal Bronco cut wire is shown in Figure 8.3.

#### 8.3.3.1 Properties of Wire Electrode (Bronco Cut Wire)

We used an alloy of copper and zinc in which the percentage of zinc will range from 40 to 53%. If zinc has such fantastic flushability (it is better for cadmium and magnesium, but cadmium is a poisonous metal and magnesium is an extremely transition metal), the ultimate wire will be created by a pure zinc coating. In theory, possibly,

**FIGURE 8.3** Cryogenic-treated and normal Bronco cut wire.

**TABLE 8.2**
**Specifications of Bronco Cut Wire**

| Core | Cu |
|---|---|
| Coating | CuZn50 |
| Tensile strength | 520 MPa/75,500 PSI |
| Elongation | 1% |
| Colour | Brown |

but it does not really turn out that way in fact. Since zinc has a low fusing temperature and is only positioned on the central wire surface, the spark discharge force appears to blow up the zinc away from the wire core surface until it has an opportunity to survive up to its maximum capacity.

Therefore, we require a coating with a higher zinc concentration and a reasonably large melting temperature, which will contribute to the core wire being well adhered to. Any of these things can be done (at 0.9-mm diameter) by heat treating the zinc-coated wire. This method is termed as diffusion annealing. Diffusion will occur under proper conditions at a high temperature, even in an inert gas surrounding. Diffusion is the mechanism by which atoms diffuse from high-concentration regions to lower-concentration areas. In our situation, the atoms of zinc diffuse into brass, and the atoms of copper diffuse into zinc from brass. The zinc coating converts this diffusion process into a high-brass zinc alloy that is rich in zinc and has a comparatively large melting point and is metallurgically bound to the core material. The specification of Bronco cut wire is given in Table 8.2.

### 8.3.4 Selection of Process Parameters and Levels

The very basic aim of optimization experiment is to determine the most influencing parameters on desired output. In WEDM, there are operating parameters such as pulse-on time ($T_{on}$), pulse-off time ($T_{off}$), current (I), voltage (V), wire tension (WT), dielectric pressure and type of wire. Operating parameters are selected by observing the literature survey trends or by doing trial and error. By doing these, type of wire and workpiece (normal and cryogenic-treated), $T_{on}$, $T_{off}$, I, V and WT have been selected for this work. The output parameter is selected as SR. Table 8.3

**TABLE 8.3**
**Process Parameters and Levels**

| Sr.no | Parameters | Unit | Level 1 | Level 2 | Level 3 |
|---|---|---|---|---|---|
| 1 | Workpiece and wire | – | Normal | Cryogenic-treated | |
| 2 | Pulse-on time ($T_{on}$) | μs | 105 | 110 | 115 |
| 3 | Pulse-off time ($T_{off}$) | μs | 40 | 45 | 50 |
| 4 | Current (I) | A | 100 | 110 | 120 |
| 5 | Servo voltage (SV) | V | 15 | 30 | 45 |
| 6 | Wire tension (WT) | N | 2 | 3 | 4 |

shows the levels of operating parameters that have been selected by trial-and-error method.

### 8.3.5 Selection of Orthogonal Array

In the present study, according to the Taguchi method an $L_{18}$ ($2^1 \times 3^5$) orthogonal matrix is being selected for the experimentation. The reason to select this Taguchi orthogonal array is to accommodate the levels. We have one factor, namely type of wire and workpiece (normal or cryogenic-treated), that can only create two levels. $T_{on}$, $T_{off}$, I, V and WT are other variables that we have. In order to achieve the best results with a minimal number of experiments, the Taguchi approach uses a technique that applies orthogonal matrix of statically constructed experiments, thereby minimizing time and expense of testing. For the determination of output characteristics, the signal-to-noise ratio (S/N) is used in the Taguchi process. The S/N can be interpreted as per the results using higher is better, smaller is better and optimal methods are trivial. The experimental design and results of SR are given in Table 8.4.

### 8.3.6 Tested Specimen

The specimens are cut according to the orthogonal table and process parameters (Table 8.4). The cut specimens are as shown in Figure 8.4.

**TABLE 8.4**
**Experimental Design and Results**

| Run No. | Process Parameters | | | | | | SR (µm) |
|---|---|---|---|---|---|---|---|
| | Wire and Workpiece | Ip (amp) | $T_{on}$ (µs) | $T_{off}$ (µs) | V (V) | Wire Tension (N) | |
| 1. | Normal | 100 | 105 | 40 | 15 | 2 | 2.518 |
| 2. | Normal | 100 | 110 | 45 | 30 | 3 | 2.994 |
| 3. | Normal | 100 | 115 | 50 | 45 | 4 | 3.397 |
| 4. | Normal | 110 | 105 | 40 | 30 | 3 | 1.7126 |
| 5. | Normal | 110 | 110 | 45 | 45 | 4 | 2.821 |
| 6. | Normal | 110 | 115 | 50 | 15 | 2 | 3.8 |
| 7. | Normal | 120 | 105 | 45 | 15 | 4 | 3.2793 |
| 8. | Normal | 120 | 110 | 50 | 30 | 2 | 3.6433 |
| 9. | Normal | 120 | 115 | 40 | 45 | 3 | 1.777 |
| 10. | Cryogenic-treated | 100 | 105 | 50 | 45 | 3 | 1.379 |
| 11. | Cryogenic-treated | 100 | 110 | 40 | 15 | 4 | 3.6526 |
| 12. | Cryogenic-treated | 100 | 115 | 45 | 30 | 2 | 3.4246 |
| 13. | Cryogenic-treated | 110 | 105 | 45 | 45 | 2 | 1.3623 |
| 14. | Cryogenic-treated | 110 | 110 | 50 | 15 | 3 | 2.9136 |
| 15. | Cryogenic-treated | 110 | 115 | 40 | 30 | 4 | 3.717 |
| 16. | Cryogenic-treated | 120 | 105 | 50 | 30 | 4 | 1.5136 |
| 17. | Cryogenic-treated | 120 | 110 | 40 | 45 | 2 | 3.6733 |
| 18. | Cryogenic-treated | 120 | 115 | 45 | 15 | 3 | 3.8503 |

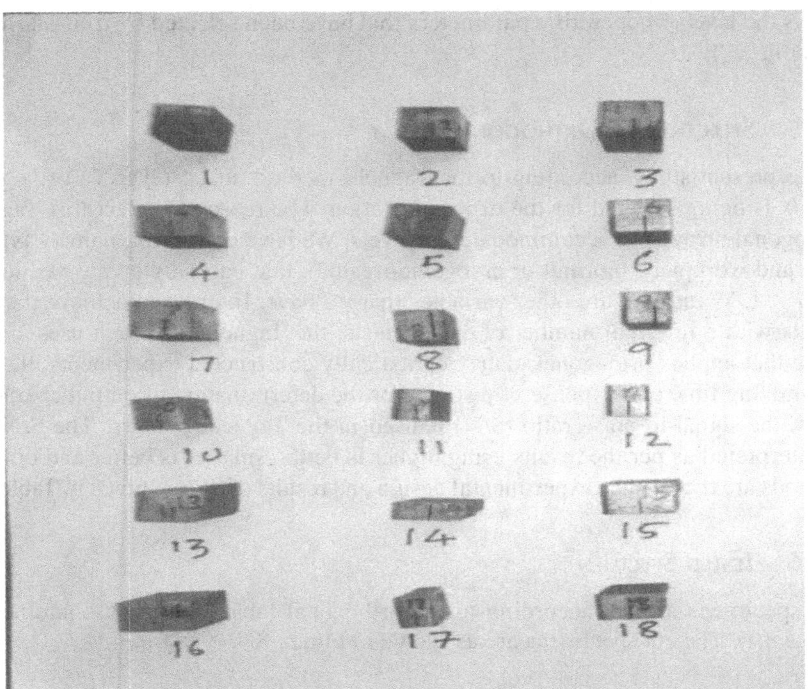


**FIGURE 8.4** Machined specimens.

## 8.4 RESULTS AND DISCUSSION

### 8.4.1 Impact on Surface Roughness (SR)

Experiments were performed using the $L_{18}$ orthogonal array to assess the impact of operating parameters on the SR. The experimentally obtained SR data and their corresponding S/N are shown in Table 8.5 and the mean of S/N and SR values are shown in Figures 8.5 and 8.6, respectively, for each vector at levels 1, 2 and 3.

Figures 8.5 and 8.6 show that cryogenic therapy has a beneficial effect on SR. For an increase in pulse-on time, SR increases; and with an increase in pulse-off time, SR declines. It is evident from the main effect curves that the measured value of SR decreases first and then increases with a further increase in wire stress. It is also apparent from the statistic that SR continues to decrease as servo voltage increases. But the SR decreases first in the case of current and then increases significantly in the third stage.

From main effect plot for SR, the optimum parameters for increase in SR are as follows. Wire and workpiece are selected as level 1, which are cryogenic-treated, current (110 A) is selected as level 2, pulse-on time (105 µs) is selected as level 1, pulse-off time (45 µs) is selected as level 2, servo voltage (45 V) is selected as level 3 and wire tension (3 N) is selected as level 2.

**TABLE 8.5**
**S/N for SR per Run**

| SR | S/N |
|---|---|
| 2.518 | –8.0211 |
| 2.994 | –9.5250 |
| 3.397 | –10.621 |
| 1.7126 | –4.6731 |
| 2.821 | –9.0080 |
| 3.8 | –11.595 |
| 3.2793 | –10.315 |
| 3.6433 | –11.229 |
| 1.777 | –4.9937 |
| 1.379 | –2.7912 |
| 3.6526 | –11.252 |
| 3.4246 | –10.692 |
| 1.3623 | –2.6854 |
| 2.9136 | –9.2886 |
| 3.717 | –11.403 |
| 1.5136 | –3.6002 |
| 3.6733 | –11.301 |
| 3.8503 | –11.709 |

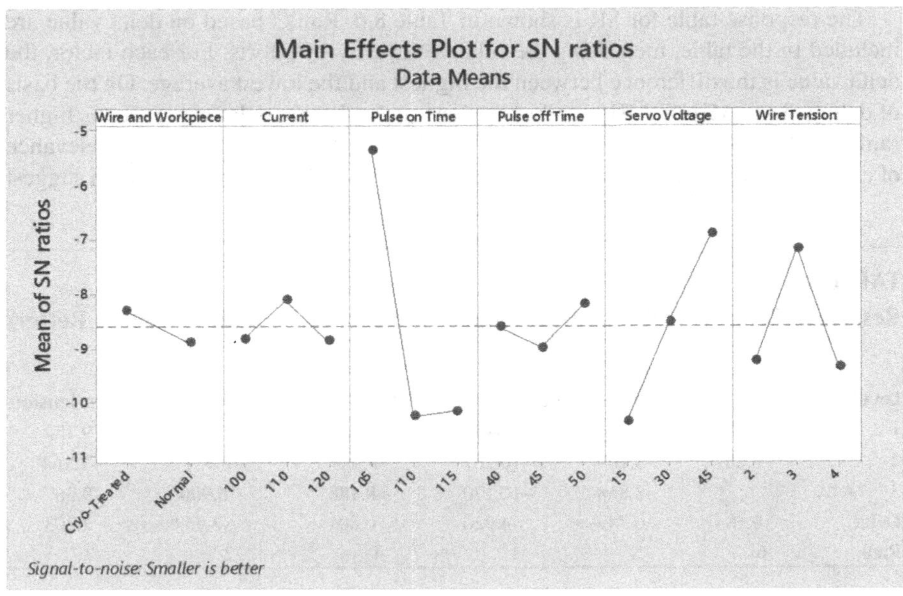


**FIGURE 8.5** Main effect plot for S/N of SR.

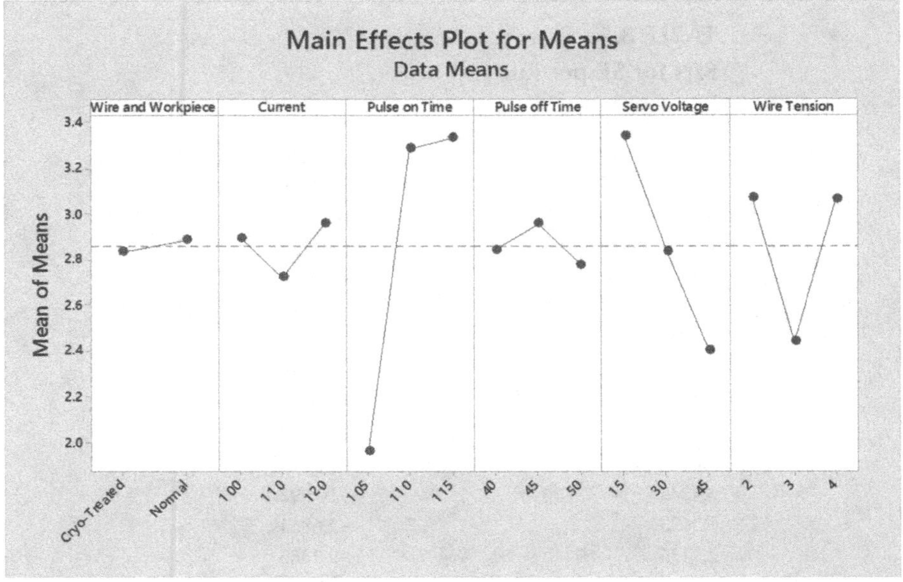


**FIGURE 8.6** Main effect plot for means of SR.

For these parameters, as mentioned above, SR is found to be optimum at 1.332 μm. This value is optimum if results of the Taguchi orthogonal array are compared. Smaller the SR, better the performance characteristic.

The response table for SR is shown in Table 8.6. Ranks based on delta value are included in the table, measuring the relative severity of efforts. For each factor, the delta value is the difference between the highest and the lowest average. On the basis of delta values, MINITAB18 toolbox assigns ranks; higher delta value means higher ranks, and low delta value means low rank. The ranks reflect the relative relevance of each element to the analyzed response. The measured ranks and delta data suggest

**TABLE 8.6**
**Response Table for SR (Means of Signal to Noise (S/N) Ratio Larger Is Better)**

| Level | Wire and Workpiece | Current | Pulse-on Time | Pulse-off Time | Servo Voltage | Wire Tension |
|---|---|---|---|---|---|---|
| 1 | **−8.303*** | −8.817 | **−5.348*** | −8.608 | −10.364 | −9.254 |
| 2 | −8.887 | **−8.109*** | −10.267 | −8.989 | −8.521 | **−7.164*** |
| 3 | | −8.858 | −10.170 | **−8.188*** | **−6.900*** | −9.367 |
| Delta | 0.584 | 0.749 | 4.920 | 0.801 | 3.464 | 2.203 |
| Rank | 6 | 5 | 1 | 4 | 2 | 3 |

*(*=Larger Numerical Value of Process Parameter)*

**TABLE 8.7**
**ANOVA for Surface Roughness**

| Source | DF | Seq SS | Contribution | Adj SS | Adj MS | F-Value | P-Value |
|---|---|---|---|---|---|---|---|
| Wire and workpiece | 1 | 1.537 | 0.84% | 1.537 | 1.5368 | 0.33 | 0.588 |
| Current | 2 | 2.129 | 1.16% | 2.129 | 1.0646 | 0.23 | 0.803 |
| Pulse-on time | 2 | 94.924 | 51.83% | 94.924 | 47.4618 | 10.13 | 0.012 |
| Pulse-off time | 2 | 1.928 | 1.05% | 1.928 | 0.9642 | 0.21 | 0.819 |
| Servo voltage | 2 | 36.038 | 19.68% | 36.038 | 18.0191 | 3.85 | 0.084 |
| Wire tension | 2 | 18.476 | 10.09% | 18.476 | 9.2381 | 1.97 | 0.220 |
| Error | 6 | 28.106 | 15.35% | 28.106 | 4.6844 | | |
| Total | 17 | 183.139 | 100.00% | | | | |

that pulse-on time has the biggest effect on the MRR and those parameters such as wire stress, pulse-off time, form of wire and workpiece, servo voltage and current in their order of significance are followed.

The ANOVA findings are seen in Table 8.7, and it is obvious from the table that pulse-on time is the key factor affecting (contributing 51.83% to output measurements), followed by servo voltage (19.68%), wire tension (10.09%), current (contributing 1.16%), pulse-off time (contributing 1.05%) and wire and workpiece (0.84%).

## 8.5 CONCLUSION

The following can be concluded from our experiment:

1. The surface roughness of machined specimens minimized from 2.518 μm to 1.332 μm, that is 52.899%.
2. From ANOVA results, it is clear that pulse-on time contributed more, i.e. 51.83%.
3. The cryogenic-treated workpiece shows minimum SR at pulse-on time = 105 μs, pulse-off time = 45 μs, servo voltage = 15 V and wire feed =3 m/min.

## REFERENCES

1. Bijaya Bijeta Nayak and Siba Sankar Mahapatra, "Optimization of WEDM process parameters using deep cryo-treated Inconel 718 as work material," *Engineering Science and Technology, an International Journal* 19 (2016) 161–170.
2. M. M. Dhobe, I. K. Chopde and C. L. Gogte, "Optimization of wire electro discharge machining parameters for improving surface finish of cryo-treated tool steel using DOE," *Materials and Manufacturing Processes* 29 (2014) 1381–1386.
3. Jatinder Kapoor, Sehijpal Singh and Jaimal Singh Khamba, "Effect of cryogenic treated brass wire electrode on material removal rate in wire electrical discharge machining," Proceedings of the Institution of Mechanical Engineers, Part C: *Journal of Mechanical Engineering Science* 226 (2012) 2750–2758.
4. G. Ugrasen, M. R. Bhagawan Singh and H. V. Ravindra, "Optimization of process parameters for SS304 in wire electrical discharge machining using Taguchi technique," *Materials Today Proceedings* 5 (2018) 2877–2883.

5. Vinod Kumar, Neeraj Sharma, Kamal Kumar and Rajesh Khanna, "Surface modification of WC-Co alloy using Al and Si powder through WEDM: A thermal erosion process," *Particulate Science and Technology* 36 (2017) 878–886. DOI: 10.1080/02726351.2017.1317308.
6. Amitesh Goswami and Jatinder Kumar, "Trim cut machining and surface integrity analysis of Nimonic-80A alloy using wire cut EDM," *Engineering Science and Technology, an International Journal* 20 (2017) 175–185.
7. Ashish Goyal, "Investigation of material removal rate and surface roughness during WEDM of Inconel 625 super alloy by cryogenic treated tool electrode," *Journal of King Saud University-Science* 29 (2017) 528–535.
8. U. A. Dabade and S. S. Karidkar, "Analysis of response variable in WEDM of Inconel 718 using Taguchi technique," *Procedia CIRP* 41 (2016) 886–891.
9. Brajesh Kumar Lodhi and Sanjay Agarwal, "Optimization of machining parameters in WEDM of AISI D3 steel using Taguchi technique," *Procedia CIRP* 14 (2014) 194–199.
10. Shivkant Tilekar, Sankha Shuvra Das and P. K. Patowari, "Process parameter optimization of wire EDM on aluminium and mild steel by using Taguchi method," *Procedia Materials Science* 5 (2014) 2577–2584.
11. S. S. Mahapatra and Amar Patnaik, "Optimization of wire electrical discharge machining (WEDM) process parameters using Taguchi method," *International Journal of Advanced Manufacturing Technology* 34 (2007) 911–925.
12. Ko-Ta Chiang and Fu-Ping Chang, "Optimization of the WEDM process of particle reinforced material with multiple performance characteristics using grey relational analysis," *Journal of Materials Proceesing Technology* 180 (2006) 96–101.
13. Anil Kumar Singla, Jagtar Singh and Vishal Sharma, "Processing of materials at cryogenic temperature and its implications in manufacturing: A review," *Materials and Manufacturing Processes* 33 (2018) 1603–1640. DOI: 10.1080/10426914.2018.1424908.
14. D. N. Korade, K. V. Ramana, K. R. Jagtap and N. B. Dhokey, "Effect of deep cryogenic treatment on tribological behaviour of D2 tool steel - An experimental investigation," Materials Today: Proceedings 4 (2017) 7665–7673.
15. Paolo Baldissera and Cristiana Delprete, "Deep cryogenic treatment: a bibliographic review," *The Open Mechanical Engineering Journal* 2 (2008) 1–11. DOI: 10.2174/1874155X00802010001.
16. Debdulal Das, Apurba Kishore Dutta and Kalyan Kumar Ray, "Sub-zero treatments of AISI D2 steel: Part II. Wear behavior," *Materials Science and Engineering* A 527 (2010) 2194–2206.

# Index

## N



## O



## P



## R



## S



## T



## U



## V



## W

Printed in the United States
by Baker & Taylor Publisher Services